HOLOGRAPHY MARKETPLACE

Fourth Edition

The Reference Text & Sourcebook

for Holography Worldwide

The original version of this book contained holograms from the vendors and many of them are no longer in business. Therefore this version of the book contains everything that was in the original version but it has no holograms.

edited by Franz Ross and Brian Kluepfel

CIP Information

Holography Market Place - Fourth edition.

Bibliography:
Includes index.
1. Holography.
2. Holography industry
Directories.

ISBN 978-0-89496-097-0
©1993 by Ross Books

For information about purchasing mailing labels of the database, please write to the address below.

PREFACE

The Holography Market Place is a publication which provides an overview of the holography industry, detailed information on holography, and a database of holography-related businesses.

In compiling our database we accommodated all questionnaires returned by our deadline and made extensive phone calls to ensure the accuracy of our addresses. Nearly all of all businesses listed in this edition were either telephoned or faxed, mailed in questionnaires, or otherwise contacted to verify their listings.

Our aim is to list businesses whose primary area of activity directly involves holography. In fairness, we try to list only the main headquarters of each business; if it is an international business we list main offices for each country applicable. Separate listings for individuals and businesses were incorporated into one comprehensive listing.

If your business is new or did not receive our questionnaire, please mail us information about your company on your letterhead to the address below.

All ($) signs represent US dollars, and (£) represent British pounds sterling. A comma is used to separate thousands (i.e. 2,000 means two thousand as opposed to 2.000).

We realize we are not infallible in our knowledge of the industry. If you feel we have omitted anything or made any factual errors, please write us. We do read the mail and will make every effort to improve and expand in subsequent editions. Address all correspondence to:

Editor, HMP
P.O. Box 4340
Berkeley, CA 94704 USA
Telephone: (1) (510) 8412474
FAX: (1) (510) 841 2695

About the cover:

The photopolymer hologram on the cover is an example of Polaroid's "Mirage" process. The following description has been provided by Polaroid:

Working from your original artwork, model or object, we will supervise the step-by-step production of your hologram. Virtually anything can be used as reference, including sculptures, toys, watches, small objects, scale models and of course two-dimensional artwork, drawings or photographs. The important thing to remember is that the holographic image is never enlarged or reduced. It must always appear the same size as the original. From there our artists, designers and engineers will control the entire production process, showing you proofs, and gaining your approval, throughout the origination and manufacturing cycle which takes 8 to 10 weeks.

1. The process begins with your supplying the camera ready artwork or other 1:1 source material. We will consult with you closely to determine its effectiveness as final holographic art.

2. A three-dimensional representation of your artwork is produced by our team of modelers.

3. Holographic mastering then takes place and a proof sample is submitted for your approval. While representative of the final hologram, this sample will vary somewhat from the final in terms of color and brightness.

4. Final production (5-6 weeks). Once we have your approval on the hand-made sample, we prepare your hologram for final production. Mirage photopolymer film is ganged and mastered. The images are then laminated with pressure-sensitive adhesive and die cut to specifications.

Specifications: Original source: 1:1 models, sculptures, flat art, photography multiplex, computer-generated or pulsed images.

Image size: Mirage holograms can be as large as 10" x 14" or as small as 0.5 square inches.

Applications: Machine - or Hand-applied
(Mirage holograms require no hot stamping)

Our custom and stock holograms have produced phenomenal marketing successes for such companies as AT&T, IBM, G.D. Searle, and 20th Century Fox. Whatever your communication needs--from advertising, to direct mail, packaging to point-of-purchase--Polaroid's Mirage holograms hold the perfect image.

Polaroid Corporation: ☎(800) 237-5519 or
(617) 577-4307

Table of Contents

Table of Contents

Table of Contects

Table of Contects

Chapter 12

Recording Materials 103

Table of Contents

Index to Advertisers

Introduction

Statement of Purpose

Welcome to the fourth edition of Holography Marketplace. This publication is designed to provide an overview of the holography industry, as well as a database of businesses for those interested in commercial contacts. We describe in some detail how holography is used by different professionals so that business people interested in getting involved in holography can gain a general understanding, and holographers in one field can examine the work of those in another field.

Holography is Unique

Holography is not only vital to some industries, but in many cases it is the only phenomenon that can produce the results we seek. For example, on the cover of this book is a hologram of a model of San Francisco's famous pyramid- shaped TransAmerica Building. There is a beam of light projecting from the top of the building toward you, the viewer.

If you allow your eyes to focus on the beam, and then take your finger and touch the end of the beam, you will find that it is focusing in space about 7.5 cm (3") above the cover. We all know that there are many ways to produce tricks of optical illusion, but the amazing thing about holography is that this is not a trick-- the hologram is actually forming an image in space, and holography is the only proven method for accomplishing this feat.

How the Book is Designed

This is the fourth edition of this publication. We have taken many of the topics covered in the past editions and re-organized them into the following sections:

Selling Holography to the Public

In past issues we have not devoted a section to this topic, but from now on we will. As the title suggests, this chapter is devoted to the selling holograms directly to the public through shops, direct mail, etc.

Holography Fundamentals

In this section, we discuss the principles of how holograms are made, and provide a description of the different varieties.

Holographic Applications

This explains more in detail how embossed holograms, CGH, HOEs, etc. are made.

Equipment

This area is devoted to recording materials and lasers.

In many cases new material was contributed by professionals involved in a particular field.

We welcome comments and questions. We do read the mail, and we take seriously your suggestions in creating each new edition.

SECTION 1

Selling
Holography
to
the
Public

1

Sales & Distribution

This chapter talks about how holography shops operate, and the chain of distribution which brings the final product from a holographer's studio to the final consumer. The last part of this discussion focuses on developing public awareness of fine art holograms.

Holography is unquestionably a business for the Nineties. Still in its infancy, it has an annual growth rate that is better than many long-established enterprises. The positive aspect of this growing field is that there is not a competitor waiting on every corner. Another advantage of this market is that while many mature businesses cringe at the thought of advancing technology destroying their markets, holography thrives on advancing technology.

Holography Shops

There are a number of successful shops starting up that deal primarily in the sale of artistic holograms. Furthermore, some of the successful shops have started small chains and are willing to make deals with people wanting to start similar shops. In effect, we are probably seeing the birth of small franchises, although there are no formal franchises.

Start-Up Questions ...

Anyone interested in this kind of venture immediately has a number of questions. Among them are:

1) Money: How much capital do I need?
2) Location/demographics: Where should I open a shop? Who are the typical customers?
3)What sells: What kind of items do I need to sell to ensure a steady monthly income?

4) *Consultant needed:* Do I need a consultant, and where can I get the cheapest and best advice?
5) Discounts/payments: What kinds of discounts and payment terms can I expect from wholesalers? Are these arrangements exclusive (can I buy from more than one wholesaler?)

... Experienced Answers

We questioned some established retailers and wholesalers and got the following feedback:

1) Money: The amount of capital you need depends on how large an operation you want to have. On the low end, you can have a "cart" operation, as opposed to an actual shop, for as little as $5,000. Buying starting stock for a small shop of about 300 to 500 square feet will cost $15,000 or more (depending on what you want to buy). The optimal size shop is about 800 to 1000 square feet and a good figure to start with is $40,000 for merchandise. You should figure another $10,000 to $12,000 in starting charges to buy lighting and showcases.

2) Location/demographics: This is probably the most important decision of all, and everyone polled said a "changing buyer population" is the most desirable aspect of a location. It was also pointed out that malls are never as good as tourist spots. The bottom line, in the words of one respondent, is that you "really need new buyers constantly." Customer profile is over 25 and under 50 years of age; from high socioeconomic brackets. This is still primarily a gift item business so we have heavy sales at Christmas-time and lots of credit card purchases.

3) *What sells?:* This can change depending on the type of shop you are setting up. An "art gallery" type of business can sell most of its trade in items costing over $100, but most shops do 80% of their business in the "over $20 but under $50" range. An assortment of items in this range, including framed or matted holograms and watches is recommended. Some low-end items (under $20) should be on hand for the impulse buyer that just wants to have a inexpensive conversation piece.

4) *Consultant/franchise needed?:* We are at the birth of this industry, and the great advantage is that you have a relatively unique, or at least uncommon and different, product to sell. There are not yet huge "McDonald's" type of franchises. What is happening right now in your favor is that wholesalers really want your business if you are going to open a shop. You represent a steady cash flow to them, and business is business. More than one of them that we spoke to would be willing to come to your place of business, help set up lighting, etc. with you and give you their experience if you order from them and pay their travel plus time.

5) *Discounts/payments:* Since this is a very new business the kinds of deals you can make vary greatly. In general, though, this industry follows trade practices used in the gift industry. You are generally told by the wholesaler what the item will cost you, the retailer, to buy. You are also told a "suggested retail price" and you are then left to mark the item up to whatever price you feel is right for your market. As with many industries, a new customer may have to pay cash for the first order or two, and then establish terms. There are promotions from time to time where some wholesaler may offer, for example, a large discount on a display case if so much merchandise is bought. You need to call different wholesalers and compare.

The Chain of Distribution

So that you can get an idea of how this market functions, we will go through the chain of distribution, from the creator of the hologram to the final buyer. You should understand, of course, that this is a young industry and these are approximations, but they are based on discussions with businesses in the industry.

Copyright Holder

The distribution process starts with the copyright holder. As its name implies, copyright means that no one has the right to copy a unique work of art made by someone else without that person's permission. In the case of holography, the original work of art is either a model or a computer program that generates a hologram. The copyright holder has the right to license or sell their work of art, and this is usually what happens.

Whoever creates the unique work of art that later becomes a hologram is considered the copyright holder of the image. It is also possible to copyright a completed hologram as a work of art, provided the objects in the hologram are not already copyrighted by someone else. There are statutory limits stating that after a number of years, a piece of art is in the public domain and anyone may use the image.

All works of art can be copyrighted, and for any unique work of art there exists only one copyright holder. Of course, the copyright holder may be a group of people such as a business. Every hologram, if properly copyrighted, has only one owner, the copyright holder. Holograms that are not copyrighted can be copied by whoever has the master. This person can make as many copies as he wishes.

Most commonly, a business pays an artist to make a model and the artist then turns over all copyright privileges to the business. This is what is known in the legal trade as "work for hire", and it must be done in writing. If would be a mistake to ignore documenting who owns the copyright to the original work of art.

Sometimes the copyright holder is a business that pays a manufacturer to make a master hologram and multiple copies, and sometimes the manufacturer is the copyright holder and can then make masters and copies itself. In order to keep our discussion as clear as possible, we describe each of the businesses in our distribution chain as separate entities. Therefore, the copyright holder is at the top of the chain and controls all distribution.

The Manufacturer

The manufacturer is only a manufacturer and gives the buyer the best price for creating multiple copies of the hologram in question. The manufacturer's price is important because the copyright holder estimates the retail price of the product based on the unit cost from the manufacturer. To get the unit cost, take the total bill from the manufacturer and divide it by the number of copies made.

Everything depends on volume. The more units you manufacture, the lower the unit manufacturing price. The lower the unit manufacturing price, the lower the retail price you can offer the product for. It is a broad generality, but many copyright holders take their unit manufacturing cost and mark it up approximately five times to get the suggested retail unit price of their product.

Selling Down the Chain

Let's assume the copyright holder decides to make a hologram. What does the distribution channel look like from this point on?

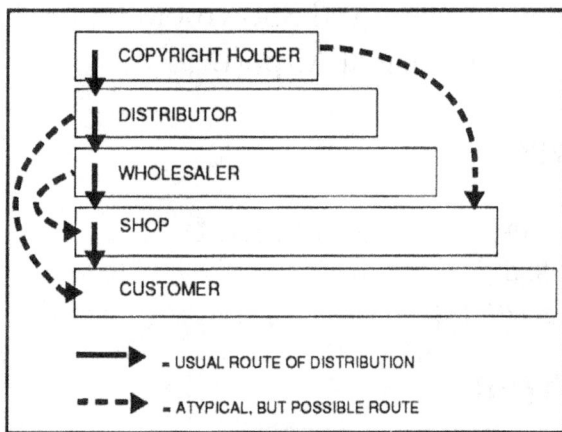

Chain of holography distribution

The levels shown above exist in the holography market and there are some rough numbers that can show, approximately, the relationship between the different levels.

It should be pointed out, however, that this is a young industry and frequently a business does several of the above functions. There exists one business, for example, that owns copyrights to some holograms that they manufacture; they import and act as a distributor for certain products; they sell products to other wholesalers; they wholesale to stores; and also have a storefront. They are everything but the customer! There are other businesses, however, that engage in only one level, like a shop or distributor. We will discuss each of the above businesses as separate entities in the distribution chain.

We feel it is our duty to give some idea of how discounts work in this industry. It is an impossible task since there is no established order of business. After some research, however, we have found a rough formula for how an item is discounted. The chart to the right is for an item costing $50.00 to the final customer. A description of each of the businesses follows.

The Distributor

The distributor buys the product in large volume, and frequently has a contract from the copyright holder that yields some kind of exclusive territorial right for a product. The copyright holder then may have the leverage to demand that the distributor not sell competing products.

The distributor generally handles all import problems, such as customs, and sometimes handles translation of written material that accompanies the product. Because there are two more discounts that have to be taken before the product reaches the customer, the distributor gets a very large discount from the copyright holder and has the obligation to order and sell large quantities of the merchandise.

The Wholesaler

The wholesaler is the link to the retailers. The essential function of a wholesaler is to get the product into shops. Wholesalers hire sales representatives or commission agents who visit shops and persuade them to buy the product. A good sales representative or agent should help the retailer set up the display, keep inventory current, and generally help to ensure that the holograms look good and sell to the consumer. It is important that the wholesaler's representative visit the retailer regularly.

A wholesaler normally carries stock from a number of different suppliers, offering a full range of holographic items to retailers. This is a convenience to the retailer, who does not have to deal with many different hologram suppliers. Frequently the wholesaler has two catalogues. One is used for shops and one is used for direct sales to customers. Wholesalers do not generally have exclusive rights to a product.

BUSINESS	(RETAIL PRICE = US $50.00)			
	DISCOUNT OFF RETAIL PRICE	GROSS AS % OF RETAIL PRICE	COST TO BUSINESS (UNIT ITEM)	GROSS PER UNIT ITEM
COPYRIGHT HOLDER	80%	30%	$10.00	$15.00
DISTRIBUTOR	70%	10%	$15.00	$5.00
WHOLESALER	60%	10%	$20.00	$5.00
RETAILER	50%	50%	$25.00	$25.00
CUSTOMER	00%	00%	$50.00	$00.00
		100%		$50.00

Typical costs, margins, and profits for holograms

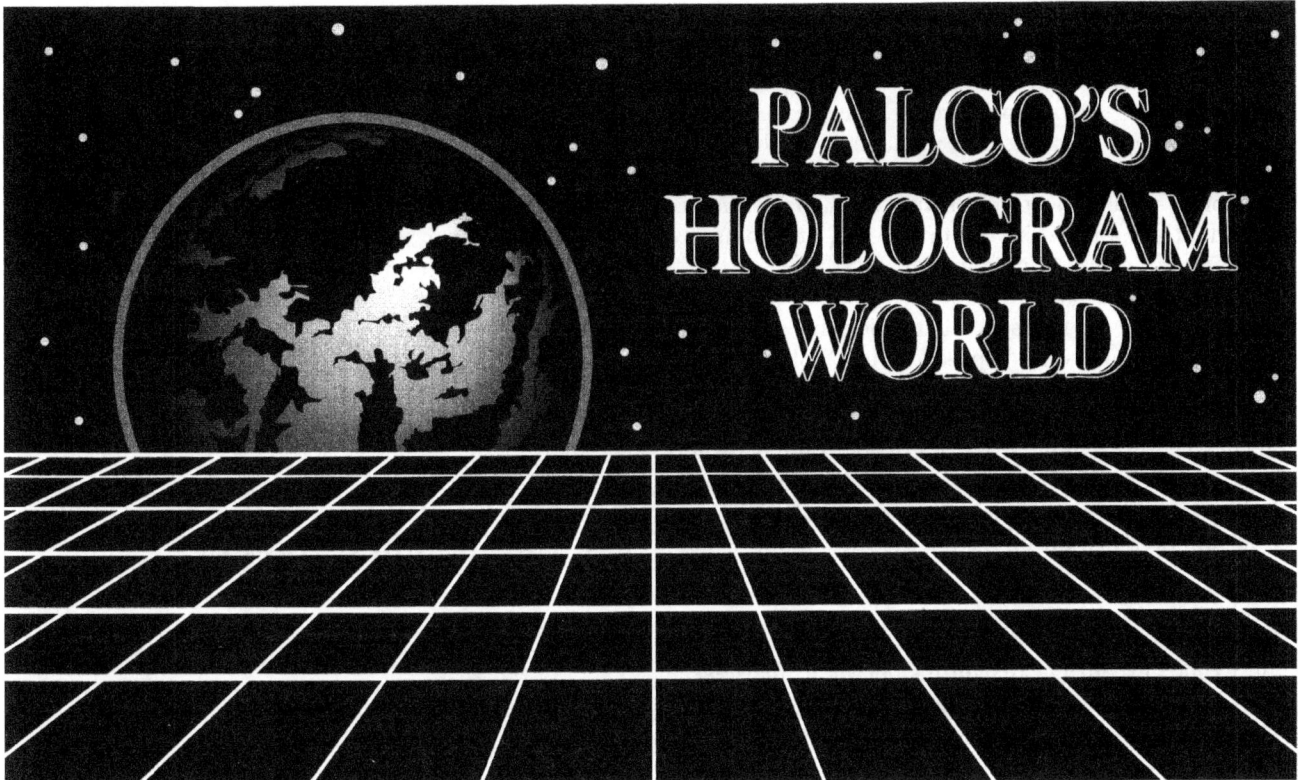

The Retailer

Retailers are the point of contact with the public, the place where the holograms are displayed, seen, and bought. Retailers of holograms include specialist hologram shops and galleries, gift shops, department stores, chain stores, stationers, jewelers and toy shops. Some are independently managed by the proprietor, and some are national chains with a central buyer.

Only the buyer in the specialist hologram shop, usually the proprietor, is predisposed to buy holograms. In all other stores the person making the buying decision has to be persuaded that the saleability and profitability of holograms (or products featuring holograms) is preferable to the other goods that are on offer to him. In sales, of course, the personal touch counts, so regular visits and product updates are an important feature of the wholesaler's function.

Discounts

The exact percentage discounts between businesses varies. In some businesses, like publishing, the final retail price of the product is established by the copyright holder and discounts follow a fairly established procedure.

Holograms are sold much like gifts in the gift trade. A retailer, for example, gets a catalogue that tells him what it costs to buy an item. The price at which the item is sold (to the final consumer) is for the retailer to decide. This is called the retail price. The wholesaler sometimes suggests that the retailer mark the item up 100% (which means that the retailer buys at an average of 50% discount off the retail price).

Wholesale Confusion

There is some necessary confusion in the industry about the word wholesaler. Retailers frequently find that the catalogues quote "wholesale prices". To the uninitiated, this implies that the prices being quoted are the prices that a distributor gives the wholesaler (wholesale prices), when actually the prices being quoted are prices at which the retailer buys the merchandise from the wholesaler.

From the distributor or wholesaler's point of view, this is an unfortunate but necessary confusion. Distributors and wholesalers point out that the retail price of a product is established by the retailer. Since wholesalers have different catalogues for retailers and end customers, they need to designate whether the prices being quoted are for the public or not. What do you call the price being quoted to retailers? You can't call it the retail price! Thus, you have to call it the wholesale price, even though it might cause some confusion.

Prices that are quoted to a distributor or wholesaler from the copyright holder are usually given as a percentage discount from the wholesale price (the price quoted to retailers).

Repackaging & Customizing

Holograms can frequently be bought by any dealer in the distribution chain in an unfinished state for a reduced price. There is a small savings that can be made if you are willing to put in the extra work, but the big appeal is that it allows you to repackage or customize the holograms for sale.

Two Items not in our Chart: Fine Art Holograms & the Independent Agency

Fine Art Holograms

The distribution chain for a fine art hologram, in theory at least, is similar although more straightforward. The day may come when holograms are a fully accepted art medium and part of the established art market; regrettably this is not true today. This means that the traditional gallery distribution system of the art world is not open to most holographic artists, nor do they have the benefit of the guiding hand of an agent, manager or dealer.

The fine art holographer's point of contact with the public is generally through exhibitions, either in specialist or public art galleries. There are several specialist galleries around the world which hold exhibitions and sell work, and many not-for-profit galleries. The specialist galleries are generally commercial operations which put a large mark-up on the artworks (or demand a large discount from the artist), whereas the not-for-profit galleries put a 15-25% mark-up on sales.

While the non-profit organizations "cost" less on any individual sale, they do not usually have continuing displays and representation for the artist. Specialists, on the other hand, do keep a permanent exhibition and often buy some of the artist's work themselves. Again, because their continued commercial existence depends on sales, the for-profit galleries tend to work hard marketing fine art holograms.

Because the market for fine art holograms is not yet fully developed, artists often sell directly to the public. Most of them know the few serious hologram collectors, and are pleased to deal with other buyers who contact them. In "dealing direct" there is often pressure on the artist to extend discounts to collectors "because there is no middle man involved."

Extending such discounts, however, undermines the growth of the distribution system and interferes with efforts to build a growing awareness of holographic art.

At last, a Database for Holographic Art Sales!

In the last year, a publication called Holart Report has come into being, with the purpose of establishing a database for the sale of holographic artwork. The market has been slow, reflecting the global economic slump. However, the publication, which began as a quarterly but will now go to a biannual or annual format, has begun a valuable service for determining fair market rates for holographic artwork.

Published by US Holographic's Gary Zellerbach, the journal has touched on important and controversial topics, such as who should get credit on the production of holograms, and which holograms should be considered "Fine Art" holograms. One important parameter set in the publication is that the works sold must be of limited editions; that is, only a certain number made, or open editions which have ceased production, therefore making them valuable. Holart Report also includes photos, when possible, of the artworks sold, with year of creation, name of artist, $ amount of sale, and number of works in original production run.

Zellerbach's publication maintains confidentiality in the reporting of sales. Sales codes determine what kind of consumer (i.e. gallery, retail store) purchased the artwork, but individual names are not printed. Hopefully this publication will do more for establishing holography as a viable entity in the fine art world, by providing benchmark values for holographic works of art.

The *Leonardo* journal (San Francisco, CA USA) published a recent edition about holographic art, wherein many artists presented their feelings on the aesthetics of holographic art, its acceptance by the mainstream, etc. If you are interested in the opinions of many of the prominent holographers on these topics, this issue is worth investigating.

Exhibitions of Holographic Art

Although we mentioned that holographic art is not readily seen everywhere, we do receive notice of ongoing or one-time exhibitions of works throughout the world. Here are some that have taken place in the recent past (we encourage you to send notices of your exhibits to Holography Marketplace):

Les seues. Cien. Jose Buitrago, Francois Mazzero, artists. October 1992. Julio Ruiz Garcia, artist. Karas Studios, Madrid, Spain.

Perpectiue, Proximities, Perceptions: Expressions in Three Dimensional Graphic and Electronic Media, scheduled for July-August 1993, Montage '93, Rochester, NY, USA

"New Directions in Holography" . Whitney Museum of American Art, November-December 1991. Artists: Rudie Berkhout, Michael Wenyon, and Susan Gamble.
ARCO '93 (Annual International Contemporary Art Fair). Crystal Pavilion, Madrid, Spain. February, 1993.

Second Annual Bay Area Holographers Exhibition, Holiday Inn, San Jose, CA February 1993. Curated by Bernadette and Ron Olson of Positive Light Holographics, Felton, CA, USA. Artists included the Olsons, Ron Hess, Nancy Gorglione.

Rdume Aus Licht, March-April 1993. Danische Zentralbibliothek, Flensburg, Germany. Artist: Frithioff Johansen

Of course, this is only a partial listing to give you an idea of the many places where holographic art has been exhibited. Consult our listings for more.

The Independent Agency

A closing word should be stated about independent agents and the market for custom-designed holograms. The work usually taken by these people are large jobs for security and advertising agencies. The markets for security and advertising holograms are similar to each other. They differ from the giftware and art markets in that the end buyer is not a consumer buying one hologram, but a company or group of companies usually buying in large quantities. The exception is the company buying only a few holograms, usually large-format, for use in promotional events such as trade shows or point-of-purchase advertising.

As discussed previously, the end-user can deal directly with the producer. Often, however, intermediary agencies bring the producer and client together. These intermediaries fill different functions from those in the giftware business. Instead of buying and selling as principals like a distributor or wholesaler does, they tend to be agents being paid either by the purchaser or the producer. They usually work on a commission in the range of 15-25%.

The intermediary ensures that the client gets a suitable hologram, well-designed, delivered on time and with the required quality. The intermediary works with client and producer to design the hologram and supervise all stages of its production, from model-making or artwork, through master and proof, to installation in its final environment. This may be hotstamping onto a credit card, insertion in a brochure or installation in a trade-show exhibition.

Everyone wants the hologram to look good - the specialist agency has to ensure that it is well-integrated into its display environment and, if it is a promotional piece, that it is designed and installed to fill its potential. This involves liaison with producers, designers, printers, print finishers and above all, the client (for more information about the interaction among clients, holographers and agencies, please read the chapter on model-making).

The hologram producer may handle all this inhouse, and many have large sales teams. As with any industry, producers find it valuable to deal directly with clients, especially large clients who bring repeat business. However, the purchaser often needs the support and advice of an independent agency.

The agency can recommend the best producer for a particular job. It can encourage a client to make the most of holograms by using large format film in a point-of-purchase application to tie in with a brochure using embossed holograms. The agency should also be aware of the many pitfalls that can delay the delivery of the order and work with the suppliers on the purchaser's behalf to ensure that problems are minimized.

If the custom hologram is for a security or anti counterfeiting application then the purchaser will want to ascertain that all stages of production are in secure facilities. Credit card companies and banks especially cannot risk holograms finding their way into the wrong hands. Several embossed hologram producers are now secure, but some have higher levels of security than others.

2

Direct Mail Marketing

Direct mail can be a cost effective way to get your message and products into the most people's hands. Here are some tips from professionals in this field interspersed with some actual experiences to guide you.

A Necessary Evil

Most businesses find that they must do some kind of direct mail advertising, whether it is using newspaper, magazine ads or fliers. If you do not advertise your business, then no one knows you exist and it is a only a short time before you are doing no business.

The object, then, is to answer this question: "What the best way to get my advertisement to my potential customers?" In this chapter we will discuss some of the dos and don'ts of the direct mail business and, hopefully, leave you with a more realistic understanding of what to expect.

The Big Sell

We have all probably seen the late night television commercials that extol the wonders of direct mail. They go something like this:

"If you want to succeed without ever leaving your office, you can. There are people who make their living through direct mail. Each day they open their mailbox and it is filled with checks, money orders, purchase orders, etc."

This seems like a relatively simple way of doing business: find people who are interested in your products, get a mailing list of such persons, and send out fliers advertising your wares. Then just sit back and wait for the dollars to roll in. Right?

Wrong! There are a number of calculations to make before doing a massive mailing and if you plan things wrong you could be eaten alive in mailing costs. Here are some formulae and facts concerning the direct mail business, and some general figures in terms of expectations.

Produce a Professional Flier

It doesn't matter how good a list you have if the flier you send in the mail is unprofessional. You might be marketing ice water to the denizens of Hell, but if your direct mail piece is lousy, your sales probably will be too.

Get a professional designer. Larger companies have an in-house designer. If you see an ad you like, call up and find out who did it. Call around the yellow page directory listings, get some price quotes, and ask to see some samples. Establishing a good working relationship with a designer could be a cornerstone of your success.

Check the ad copy. We have received direct mail pieces with major misspellings, informational inaccuracies, garbled grammar and splintered syntax. Where do they go? In the "circular file".

Read some advertising books to make sure you are using words that are proven successful, and "jump off the page" at the recipient. Make sure that someone whose grammar and spelling are trustworthy proofs the final copy before it is sent to the printer (most computer programs have a spell check feature).

Buying & Testing Your List

As a book publisher we have had several experiences in bulk mail. There are a number of companies who do nothing but sell mailing lists. Naturally, your eyes light up as you see some of these lists: "those people would definitely buy my product!"

Here's a word of caution: look before you leap. There's no way to gauge how a given audience is going to react to your mailing. Our recommendation is to "test the list": companies usually have a minimum dollar volume or number of names which you have to buy--but that doesn't mean you have to mail them all at once! You do need to mail enough so that you do not get a statistically incorrect result. Most industry wags say that you can get meaningful results with as few as 200 to 500 pieces mailed. That's enough of a number to know whether you should spend the additional time and money on postage and bundling, etc.

There are a number of companies who specialize in the selling of names. Among them are:

USA
American List Council; Princeton, NJ
  (800) 526 3973.
College Marketing Group; Winchester, MA
  (800) 677 7959;
Woodruff-Stevens, NY, NY
  (212) 685-4600;
Lakewood Lists, Minneapolis, MN
  (612) 333-0471;
List Services Corporation, Bethel, CT
  (203) 743-2600;
Direct Marketing, Pottstown, PA
  (215) 326-4966;
IBIS International Direct, NY, NY
  (212) 779-1344
Europe
Royal Mail International, London, England UK
  FAX (44) 071 3204141
European Direct Marketing Association
  Brussels, Belgium Telephone
  (32) 2 217 6309 (65-member organization of European
list brokers)

Note that in some cases you can order both labels and a duplicate copy of the labels with phone numbers of the people on the mailing list.

Seven Rules for Direct Success

Lakewood Lists offers a quarterly publication called Direct Success, which offers tips on getting the most out of your mailing efforts.

Another book on the subject is Direct Mail and Mail Order Handbook by Richard S. Hodgson, which outlines "The Seven Cardinal Rules for Direct Mail Success":

1. What is the objective?
2. Get as tightly focused a list as you can.
3. Write your copy to show what the product or service offered does for the reader.
4. Make the layout and copy fit.
5. Make it easy for the prospect to take whatever action you want him or her to take. "Direct mail is the action medium. Every mailing should call for an action of some type: inquiry, purchase, referral, contribution, phone call, visit to a local dealer--an action that the mailer wants the prospect to take. You can incite action through the appeal of the offer or through the device for responding. The appeal of the offer is of first importance. Some of the offers which overcome the barriers of human inertia and make it easy for the prospect to take action are free information, free trial offer, free gift for action, sweepstakes contests, money-back guarantee, installment terms. Devices for responding are likewise an inducement to action. Direct mail because of its very nature is made to order for responding devices. It's easy for the prospect to respond if you give him a pre-addressed stamped reply envelope or card, or a postage-free envelope or card. All other media require more effort to take action."
6. Tell your story over again. "Most mailers don't mail often enough. In many lines of business that employ salespeople, it is found most sales are completed after the fifth call. Yet scores of people try direct mail once, don't get dramatic results, and quit."
7. Research your direct mail.

Post Office

It is imperative that you use bulk mail. Any mailing of 200 or more pieces can go bulk mail at a substantial discount off the normal First Class postage rate. You will have to buy a annual bulk mail permit and have your bulk mail sorted in the order that the post office requires, but the savings, when you are putting hundreds and thousands of pieces in the mail, adds up very quickly. In the classified section of the phone book of most cities is a listing of "mailing services" or something close".) to that heading, which

includes businesses that "label, sort, tie and sack" your mail according to the post office bulk rate department specifications. Therefore, all you have to do is order your printing (fliers, etc.), have it sent to the mailing house to be labeled, sorted and made ready for the post office, and then tell them to deliver it to your post office bulk mail department.

Sales Analysis

To put you in touch with reality, we will share with you some direct mail experiences from a recent periodical. Keep in mind that there are business computer programs designed specifically to track direct mail results, which give a complete printout of sales, percentage of returns, money made versus costs, etc.

Whether you use a database or an abacus to calculate your returns, you should figure in these costs:

+ Prepress work (design, layout, typesetting)
+ Cost to buy list
+ Cost to print (paper, print and binding)
+ Cost to label, tie and sack
+ Cost for postage (to send flier)

= Thtal cost of ad

To calculate the profit:

+ Gross sales
- Royalty (if applicable)
- Manufacturing cost (raw cost) of each unit sold
- Total cost of ad (from above)
- Postage (to mail out sold units)
-Tax

= Net Profit

What to expect:

The first surge should come about two weeks after the piece is mailed. If you don't get anything by then, or response is small, this probably won't be a biggie. However, if you do get a good response in this time frame (give it up to three weeks, just to be safe) then you can expect another surge within the next two weeks after that. Then you can expect sales to trail off gradually, starting about two months after mailing, until there is insignificant activity some four months later.

Industry Averages:

It is said that if you have a targeted list and a good mailing piece you can expect a one percent re-

sponse. This figure makes it clear that you need big volume in this business. In truth, you can get zero responses, even with a very highly targeted list that usually gives an extremely good response.

Case Studies

1) **A Highly-targeted List:** Company X did a direct mailing to a small list of people, all in their specific industry, who had a proven need for the product. They received a 9.5% response. This is extraordinarily high, and it is also rare to have so highly-targeted a list.

2) **Success or Not?** A more typical successful response would be received by mailing to some subscribers to an industry magazine. In a test run, Company X mailed out 167 pieces and received 7 back for a 4.2% response. Sometimes success can be deceiving, however. This magazine had mentioned X's product in its review section, and then sent responses based on that. Of the 60,000 subscribers to the magazine, 167 took time out to ask for more information on X's product. Of those 167, 7 bought books on the first mailing. So the magazine, in effect, winnowed down the list for Company X. Had X mailed to an unqualified grouping of the magazine's subscribers, their success rate probably would not have been nearly as high. The result, means that X should start buying the list and mail to the larger audience, but should not expect quite as large a percentage response.

3) **A Dismal Failure:** X also bought a subscription list to a couple of very popular magazines they thought would have readers interested in some of their products. X learned the hard way that the more you widen the focus of your audience, the less chance of success you have. Of 1,000 fliers mailed out, only 10 were returned for a 1% response. That wouldn't be bad if you are selling $2,000.00 copier machines. However, you should make sure that the unit cost of sales is enough to recoup the printing, postage and other costs involved. Write this one off as a bad experience.

Please note that the higher the unit price of the product you sell, the fewer sales you have to make to break even. However, the higher unit price also means fewer returns on your ad.

Magazines and periodicals:

You can begin to see why people advertise in periodicals. The cost to get their advertisement into the hands of readers is far less, the publication probably stays around longer and there is less fuss in getting the job done. You also have the user actually looking at the ad longer because they are generally reading the publication from time to time. The drawback is that you are sharing space with others.

Summary

For most businesses some form of advertising is not only necessary but critical. In most cases your sales plan should include a combination of approaches. Buy some "test" mailing labels. In the very highly targeted "center-of-the-bulls-eye" category, call the people as well. Do some space advertising in publications that are targeted at your product and try to spread these projects over time so that you always have some cash flow.

References:

Direct Success, newsletter from Lakewood Publications, a division of McLean Hunter Publishing, Minneapolis, MN USA.

Marketing Without Frontiers, Second Edition. by Royal Mail International, London, England.

3
Model Making

Nearly all holograms are made from a model. Creating a model is arguably the most important step in the entire process. Veteran model-maker George Sivy talks about establishing good relations among the client, broker and holographer during the hologram's creation so that everyone is happy in the end.

A Interview With A Holography Model Maker

Although there are many steps in the production of a hologram, the first step in the three-dimensional imaging process is making a model. The success or failure of your project depends on how well you are able the convey to the artist your concept, and his/her skill in executing it.

George Sivy has been a holography model-maker for eight years. He worked for Polaroid during their first years in holography, making models which would be turned into famous images like "The Skull" which graced the cover of Holography Marketplace's third edition, as well as the 1991 Super Bowl ticket for the National Football League. He has run his own business, Gray Scale Studios, out of Boulder, Colorado for six years; he is also working with US Holographics of Logan, Utah on their current projects.

Sivy has three tips for those who are working with model-makers: plan in advance, remain in direct contact with the artist from start to finish, and be willing to pay for the artist's time and expertise. He talked briefly about each topic.

Do Some Research

Model making is the first step in a process which includes shooting the hologram, mass producing the hologram, and applying the hologram.

All the succeeding steps are dependent upon a good model--even if they all proceed without error, the final result can only be as good as the original model.

Generally clients are middlemen or brokers, dealing with an art director, says Sivy. Sometimes, for "political" reasons, the middlemen don't want to put the artist and client in direct contact with each other. These middlemen mayor may not be aware of the potential use of different angles, monochrome rendering, animation, stereograms, and other special effects.

"Nine out of ten of these people do not know what to expect in terms of time and other factors, which go into creating a piece, " he adds. Because of this, he says, there can often be "communication gaps, misunderstandings, and deadlines which are too tight ... there is a lack of communication and understanding of what it takes to create an image. Expectations should be more appropriately formed," he concludes.

This means the client should meet with the artist before the creation of the model begins, and explain what effect is desired (it might not be a bad idea to ask for sample holograms created from the artist's models). The client can outline what he/she wants; the holographer can explain what's possible, and how much work and time will be involved. "A hologram, under the best circumstances, is still an interpretation of the client's 2-D artwork," says Sivy.

Stay in Touch

Keeping in touch is absolutely essential, says Sivy. "It's crucial putting corporate politics aside--the guy signing the check has to be in direct contact with the artist right from the beginning," he says.

This means determining a schedule, and meeting regularly to determine whether the work is proceeding according to plan.

Things to keep in mind

Get exact specifications on what volume of space your model can occupy before starting. Remember that an embossed hologram has little depth. Discuss the material being used to create the model with the holographer.

"The general rule for embossed 3D holograms is that one inch is about as deep as you want to go before the hologram begins to 'smear'. This means the model should not be deeper than one inch maximum," says one holographer. "This is because embossed holograms are designed to be viewed under most lighting conditions." The final printed color of the model does matter. Colors that turn out well depend on the laser used to expose the object.

Once your model is made and painted, you can view it under the light of the laser that will make the master. This will give you a rough idea of what areas of your object the laser light reflects strongly from, and from what areas it reflects weakly. You should remember, though, that your hologram may be viewed in unusual lighting conditions and the actual color seen depends on the ambient lighting conditions.

Some of the leading embossers use unpainted models of special white plaster. It is suggested by one holographer that if you insist on painting your model, it is easier if the model maker uses a safelight (available from Roscoe) covered with a non-transparent gel filter of approximately the right color to simulate laser light. This is a simple method of judging appropriate paint colors.

Although these steps mean that the model-making may be the most time-consuming part of the hologram production, it is time well-spent. Remember--if the model is not up to snuff, nothing else will be either.

Pay the Price

Finally, artists must be paid what they are worth. Sivy says, "people have to be aware that with special effects, animated images, and larger formats you have to pay for the level of detail you are expecting. There is a lot of equipment and knowledge involved which you can't get by reading a book. The best models have the greatest amount of detail, and time equals expense."

Sivy aired a pet peeve here. "A lot of people get sold on holography through the pitch that model making is cheap--about $500.00," he says. He notes, however, that nearly any decent model will in fact cost between $2000-$3000, while pieces like the two-channel skull cost $6,500.00 (although today the same amount of work might be worth as much as $10,000) and the Super Bowl ticket image $8,000.00. "Salesman always seem to bend things a little bit and work backwards from the client," he says. 'With special effects, you have to be willing to foot the bill for r& d, If you want a unique product, you have to be willing to fund its creation."

Materials

Sivy uses "what will solve a given problem." The materials, he says, have to fill the need of whatever model he is creating, which varies from week to week and day to day, including dinosaurs and nudes for holographic watches in Taiwan. Although some materials and combinations that he uses are proprietary, he did mention the readily-available Sculpey, a synthetic clay which can be baked. "Certain elements of a given model might be appropriate enough for Sculpey. By the same token, if there's an arm extending from a body, which is thin in comparison to the rest of the model, the arm might have to be made of something totally different." Under some formats, like Argon lasers with a long exposure time, he says, he may sculpt in Sculpey and then make a mold of the model in another material.

He emphasized, "That's why people making models have to be versed in a variety of materials."

References:

Interview with George Sivy. Gray Scale Studios, 4500 19th Street, # 294, Boulder, CO, 80304 USA (303) 442 5889 (phone/fax)

Publishing & Production Executive, July 1991; "Delving Deep Into Holography"; by Kevin Samson (Light Impressions Corporation).

SECTION 2

Holography
Fundamentals

4

Holography Basics

How are holograms made? What different kinds of holograms are there? What raw materials and equipment are needed to produce them? We answer these and other basic questions in this chapter.

An Everyday Occurrence

You wake up in the morning and go down to the kitchen. Cracking open the cereal, a bright image jumps out at you from the front of the box. You walk to your mailbox, expecting that letter from your Aunt May in Canada. Affixed in the place where a regular stamp would be is yet another of those odd three-dimensional images you just saw on your cereal box. The magazine which you subscribe to also has one of these bright, reflective pieces of foil stamped on it, with an image seemingly moving across the cover.

Still bleary-eyed, you dress and get in your car. Not paying close attention to the road at this early hour, you run a stop sign. The friendly California Highway Patrolman who pulls you over takes your license and runs it over a scanning device to make sure that it's the authentic item.

This scenario is an example of the frequency with which we see holograms in our everyday life. Rarely a month or two passes without a magazine or other publication using a hologram to attract attention at the newsstand--this has become commonplace since a hologram first appeared on National Geographic in 1984. The tickets to Presidential Inaugurals and Super Bowls have holograms hot-stamped onto their surfaces to make it tough on counterfeiters.

Holography and Photography

Before citing further holographic applications, we should explain what a hologram actually is. We can start by contrasting holography with its neighboring field, photography.

Although often compared with photography, holography is really a completely different medium with different applications. They are the same only because they both are ways of capturing an image, and, at times, similar chemical processing is used in making both items. There are, however, major differences between a photograph and a hologram.

3d, Or Not 3d?

A photograph can be made under any ordinary lighting condition but the resulting image is only two-dimensional. That is, if you move a photograph from side to side you do not see around the image - you will always see nothing but a flat image, even though many gimmicks (e.g. 3-D glasses) have been marketed over the years to trick the eye into thinking it is viewing a three dimensional image.

A hologram, however, actually is a 3-D image. With some holograms, the image actually forms in air in front of the holographic plate and you can look around the object just as you would in real life. The distance the

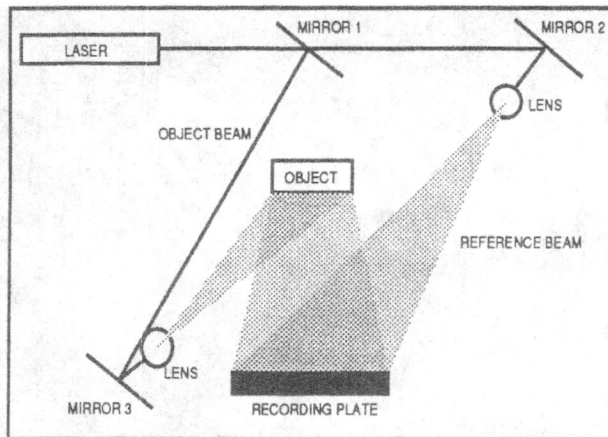

Figure 1.1. Transmission Hologram

image forms in front of or behind the holographic plate and the degree to which you can look around the object depends on how the hologram was made; there are a large variety of holograms, each with good and bad points.

Although some holograms may be viewed in ordinary daylight, they all require a very narrow, almost single, wavelength of light to be made. Because of this lasers, which can put out a single wavelength of light, are almost universally used to make holograms.

To borrow a phrase, a picture (or in this case, a hologram) is worth a thousand words. Therefore, to fully grasp what holograms look like, one should go to a local shop or gallery which displays them or order some from one of the many mail-order houses found in this publication.

Perhaps the best way to begin is by explaining, step-by-step, the making of a simple hologram. After discussing some of the fundamentals of holography, the more exotic varieties of holograms will be addressed.

Making a Transmission Hologram

Suppose one enters a studio where a simple hologram is about to be made. There is a special vibration free table in the room. On the table is a laser, some mirrors and a photosensitive plate in a plate holder. Everything on the table is arranged in a carefully measured manner. In the center of this holographic set-up is the object to be holographed.

To make the hologram, we turn on the laser. The laser beam strikes Mirror 1. Due to the fact that Mirror 1 is only partially reflective, part of the beam is reflected toward Mirror 3, and the other part passes through Mirror

1 to Mirror 2. For this reason, Mirror 1 is referred to as a beamsplitter.

The beam which passes through Mirror 1 is called the reference beam because it never actually strikes the object being holographed. After the reference beam strikes Mirror 2, it is reflected through a lens toward the photographic recording plate. The lens' function is to spread the beam so that it will cover the entire plate (in some cases, the lens is placed in front of Mirror 2; in either case its function is the same - to spread the beam).

At the same time, the other beam (which we call the object beam) reflects off Mirror 3 and passes through a lens. This lens spreads the beam out so that it illuminates the entire object. The beam then reflects off the object (hence the name object beam) and strikes the photographic recording plate.

The two beams must travel exactly the same distance so the light waves in the beams will be synchronized with each other. After exposure (exposure time depends on the laser and type of film or emulsion used), the photographic plate is developed, and the resulting developed plate is a hologram.

Holding the developed plate up to light, we see that the plate is semi-transparent. On closer inspection we see that the darkness of the plate is caused by developed emulsion. The plate seems to have countless swirls of thread-like developed emulsion which are called fringes. The fringe patterns look like the swirls that make up your fingerprints or the boundaries of topographic maps. There appears to be no order to the swirls.

Viewing the Finished Hologram

To see the completed hologram, the photographic recording plate is developed and positioned on the table in exactly the same place it was for the exposure. The object and Mirrors 1 and 3 are then removed from the table.

The laser is turned on again. When one looks through the plate, an image is seen of the original object, in its original place and at its original depth.

A detailed explanation of why this happens would occupy many pages. A simple explanation might go like this: the object and reference beam strike the photosensitive recording material at the same time and create a pattern as they expose the chemicals in the emulsion. Since both beams strike the plate at the same time, the light waves in the beams naturally interfere and combine with each other. Therefore the

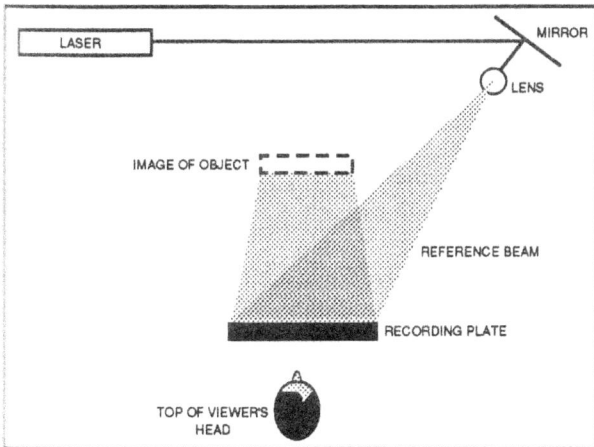

Figure 1.2. Viewing a transmission hologram

Reflection Holograms

The hologram just discussed is called a transmission hologram because the light passes through the plate to us.

It is also possible to make a hologram where the light reflects off the surface and back to our eyes for us to see the image. This is called a reflection hologram.

How are reflection holograms made? Look at Figure 1.1. If we transfer the reference beam around with mirrors so it illuminates the recording plate from the back instead of from the same side as the object beam, we create a reflection hologram, as shown below. It is that simple.

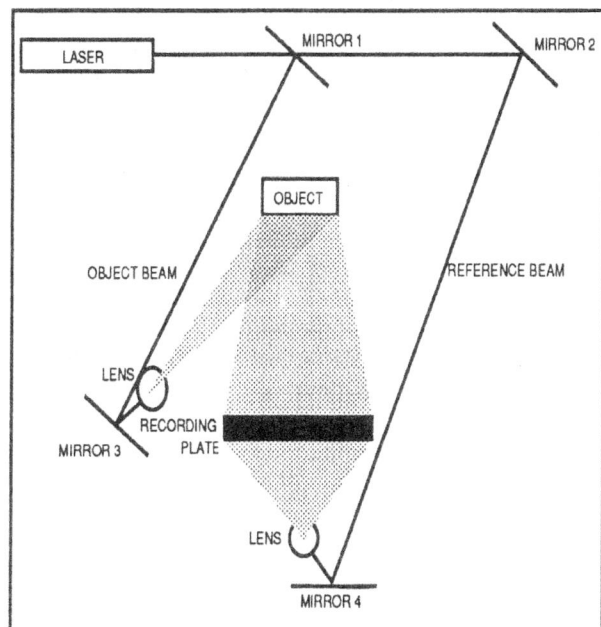

pattern we get when we develop the plate is called an interference pattern.

After development, we aim only the reference beam at the plate, at exactly the same angle that originally exposed the plate. As the reference beam goes through the plate, the interference pattern naturally causes part of the reference beam to change direction. This phenomenon is called diffraction.

The interference pattern, because it was created by light from the original object, diffracts the reference beam passing through it back in the direction of the original object. This deflected beam has exactly the same form as the beam that was originally reflected from the object, because the object beam made the patterns in the emulsion.

Now, the reference beam travels through those patterns to recreate an image of the object in the same place. In other words, a hologram mimics the way light reflects from an object, without the object being there.

In order to record a clean and clear interference pattern of our object and reference beam, we use a laser, which has a single beam of light at one wavelength. We cannot use just any light as our source, because the light from common light bulbs contains many constantly-changing wavelengths. If we make the exposure using regular light, the interference patterns would be completely blurred and useless because of the changing wavelengths and multitude of interference patterns that would be created.

As their name implies, all reflection holograms are illuminated from the same side as the viewer. In other words, when viewing a reflection hologram, light comes from your side of the plate, strikes the plate and reflects back to your eyes.

It should be pointed out that all reflection holograms can be viewed in sunlight (frequently called daylight viewable or white-light viewable holograms) whereas transmission holograms, depending on how they are made, can be either daylight-viewable or laser-viewable (only viewable with a laser light). We will discuss these categories shortly.

Within the two major divisions of holograms (reflection and transmission), there are many variations. Like any other specialized field, holography has its own lingo, and in some cases the same hologram can be described using more than one name.

We will try to clarify the sub-divisions of holograms next - but remember, each of these subsets can always be classified into the larger families of transmission or reflection holograms.

Thick and Thin Holograms

Another broad classification of holograms is made when differentiating between "thick" (sometimes called "volume") holograms or "thin" holograms. One of the reasons the words "thick" and "thin" are used in conversation is that they allow one to instantly get an idea of some of the properties of the hologram. Very thin holograms provide little depth to their object upon reconstruction. Embossed holograms, such as the images on bank charge cards, are examples of thin holograms. Thick holograms have the ability to replay or reconstruct the image with considerable depth or projection.

How do you decide if a hologram is a thick or a thin hologram? If you look closely at the surface of the hologram you see the fringes caused by the interference pattern. A hologram is considered to be "thick" if the thickness of the recording medium is considerably greater than the spacing between the interference fringes. Otherwise the hologram is considered a "thin" hologram.

The distance between interference fringes will depend on a number of things, such as the wavelength of light being used, and the density of particles in the emulsion.

These interference fringes are called Bragg planes after Sir William Henry Bragg and Sir William Lawrence Bragg, who did much of the early work on the subject. As you would expect, Bragg planes actually go all the way through the medium, but are visible to our eye only where they meet the surface.

It should be noted that in a reflection hologram, where the reference beam and object beam strike the plate from opposite sides, the Bragg planes slice through the medium at very shallow angles.

Conversely, in a transmission hologram, where the reference beam and object beam strike the plate

from the same side, the Bragg planes cut the emulsion at much sharper angles and thus are further apart.

The H-1 and Master Hologram

It is important that we cover the topic of the H-1 and master in this introduction because it is a fundamental procedure in the making of almost every commercial hologram.

H-1 stands for "hologram one", which simply means it is the first hologram you make on the path to your desired final hologram. Sometimes the H-1 is the master hologram from which you make multiple copies. Frequently, though, there is more than one hologram that needs to be made before you get the desired master hologram from which you will make copies. If this is the case, the next hologram in the sequence is called the H-2, and then H-3, and so forth.

The laser viewable transmission hologram is most often used as a master hologram. Transfer copies (making another hologram using the image on the master as the subject) can be made in quantity from the master. These transfer holograms can either be other laser-viewable transmission holograms, white-light viewable transmission holograms or reflection holograms (which are always white-light viewable).

A question that immediately comes to mind is, "Why would anyone want to make an H-2?" Well, historically one of the big problems that holographers used to have was placing the object to be holo-

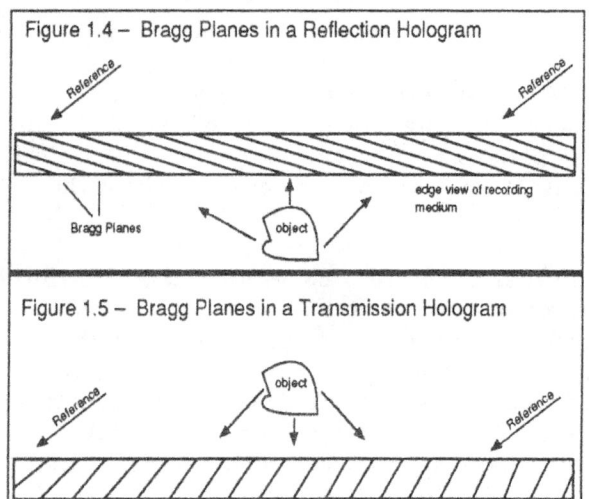

Figure 1.4 – Bragg Planes in a Reflection Hologram

Reference   Reference

Bragg Planes   object   edge view of recording medium

Figure 1.5 – Bragg Planes in a Transmission Hologram

Reference   object   Reference

graphed exactly where they wanted it. Suppose, for example, you want the object in the final hologram to appear half in front and half behind the recording plate. How would you do it? You obviously can't do it on your first shot because the object would have to be going right through your photographic plate.

This problem was solved by the following procedure:

• Make a transmission hologram. We call this H-1 because it is our first hologram.

• Since the H-1 hologram creates an image of the object, why not use the image made by our H-1 as our subject and make a hologram (H-2) of the image made by the H-1? In other words, make a hologram of our hologram. This H-2 hologram can be a transmission or reflection hologram, depending on your need.

• It sounds strange, because you are making a hologram of an image and not an object. But it works.

• Now, since you can make a hologram of the H-1's image, take time to move the image around to wherever you want it positioned. In this case, adjust the H-2 recording plate so that the image of the object is half in front and half behind the plate and then make your H-2. The problem of getting half of the object in front of the plate, and half behind, is solved.

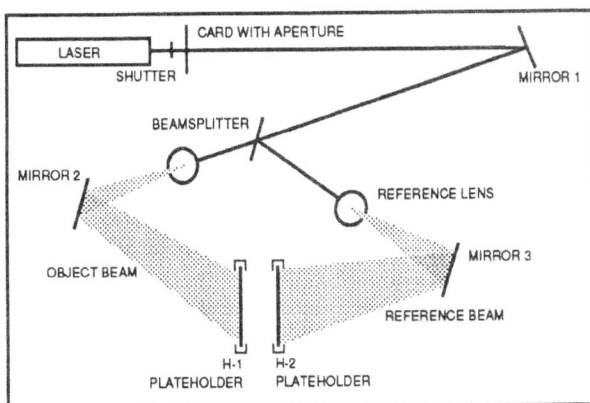

Figure 1.6 – Reflection H-2 being made from H-1

In short, there are at least three good reasons why an H-2 should be made:

1) The H-2 allows you to reposition the image of your subject. When you reposition your image from the H-1, you may make your subject focus out in front of the recording plate, behind the plate, or anywhere within the limits of your equipment (you are usually limited by the laser's ability and the quality of the optics). The creative potential here

is enormous because you are able to move solid objects around like they are ghosts. You can have two objects occupying the same space, etc. The process of moving the image around to make the H-2 is called image planing.

2) It gives the holographer a chance to brighten up the image. Since you may move your image anywhere, you can focus the image right at the recording plate. This concentrates the light directly on the recording material and brightens up the image considerably. This is commonly done in silver halide reflection holograms.

3) It saves time on remakes. If you develop the H-2 and decide you don't like the position of your subject astride the recording plate, you don't have to find the original subject and set it up again. This can be important if there are large costs in arranging the H-1 shot.

Going through the pains of making H-1, H-2, etc. to get a good master is necessary for creating most holograms. It is technically possible to get some desired holograms by skipping this process, but the results are generally very inferior. A master is almost always used for commercial jobs of value.

Sunlight-Viewable and Laser-Viewable Holograms

We mentioned earlier that although it is necessary to use a laser to make a transmission hologram, it is not always necessary to use a laser to see a transmission hologram. In fact, most transmission holograms can be seen in sunlight.

This may seem confusing, because we have said that in order to see a holographic image you have to shine the reference beam that made the hologram on the plate. This is true, but sunlight contains a multitude of wavelengths, including the one we used to make our exposure. The sun is such a great distance from earth that it appears to be a single beam of light shining on our plate. It would seem that we have only to position the plate at the proper angle, and we should see our image.

This is logical, but it also stands to reason that if sunlight passed through a transmission hologram we would also get images being formed by all of the other wavelengths that are somewhat close to the wavelength of your reference beam. These other frequencies of light would diffract at a somewhat different angle than the

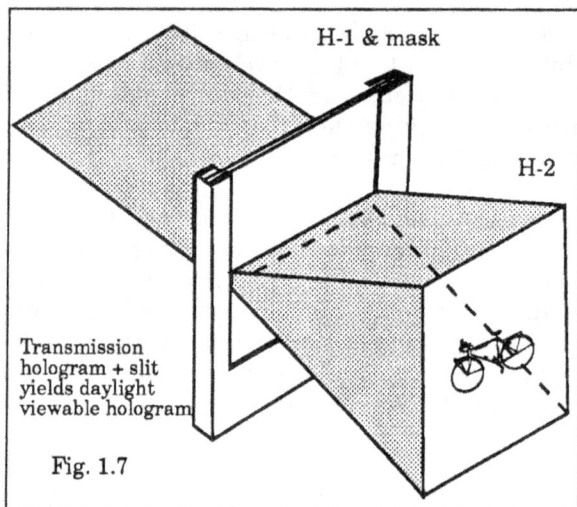

Fig. 1.7

original reference beam. The result would be a whole multitude of images forming right next to each other, creating a blur instead of a clear, crisp image.

That's exactly what does happen and it took a while for a solution to be developed. Around 1969 Dr. Steven Benton came up with a solution. The resulting hologram is sometimes referred to as a Benton hologram or, more frequently, a rainbow hologram.

Rainbow Holograms

Benton reasoned that since our problem is too much imagery at the point of reconstruction for our object, why not block off some of it? In other words, suppose we put up an opaque mask against the transmission hologram, with a long, narrow horizontal slit through which we view our transmission hologram. This would certainly clean out a lot of the annoying secondary images that are blurring the primary image's reconstruction.

This "cleaning" comes at a price, however, because the mask causes loss of vertical parallax (the ability to be able to see over and under our object). We would, however, still have our horizontal parallax (ability to see side to side around the object). Humans, with feet fixed on the ground and eyes on a horizontal plane, are actually more accustomed to horizontal parallax than vertical.

The procedure to produce this masked hologram is as follows:

1) First a normal transmission hologram is made.
2) Next, a transfer copy of the transmission hologram is made, but an opaque card with a horizontal slit in it is placed between H-1 and H-2.

If the copy hologram is viewed in the frequency of our laser light, the eyes must be positioned at the real image of the slit to see the holographic image.

Now, imagine viewing this H-2 hologram in two different colors of light. A hologram of the image made through the slit will be played back, but each of the two wavelengths of light will diffract through the hologram fringes at a slightly different angles. There will be two different images of the object, each a different color and each at a slightly different vertical position.

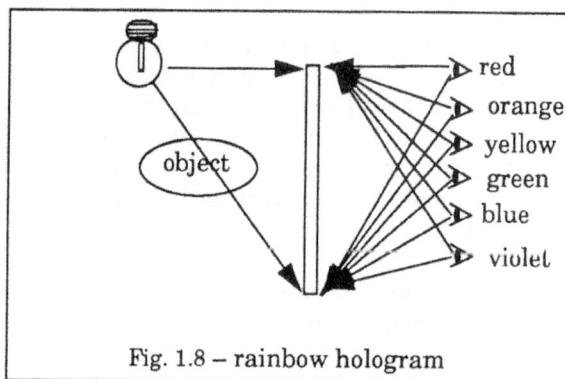

Fig. 1.8 – rainbow hologram

Next, think of the image in white light or sunlight. All of the wavelengths will reconstruct their own image, all slightly displaced vertically with respect to one another. We are faced with the same problem we had with the original transmission hologram, except the images being recreated have no vertical parallax.

As you move up and down in front of the plate the color of light will change but the image will be the same (hence the name "rainbow hologram"). As you move from side to side you will have horizontal parallax because nothing has been done to destroy it. By careful planning, the image may be made any desired color, or even a combination of colors (a multicolor rainbow hologram).

In effect, the hologram is filtering the white light, while all that is sacrificed is vertical parallax, which, as we mentioned, our two horizontally-positioned eyes usually don't miss anyway. Rainbow images are often extremely bright, because all of the frequencies in white light are being used to form the image. So the rainbow hologram technique is a way of making a transmission hologram sunlight-viewable. Other names for this are daylight-viewable or white-light viewable. They all mean the same- a hologram you can see without the need of a laser.

Some rainbow transmission holograms are displayed in art galleries on glass plates and on film. However, they are much more popular in two other forms. The two most common forms in which rainbow transmission holograms are seen are as embossed holograms and holographic stereograms.

In an embossed hologram, the light goes through a rainbow transmission hologram that has been embossed in clear plastic, strikes a mirrored backing and reflects back through the rainbow transmission hologram to the viewer's eyes. Chapter Ten of this book concerns itself entirely with embossed holography.

Holographic stereograms are made from motion picture film. There is more detail on these holograms at the end of this chapter and later in the book.

Lighting

Sunlight is not the only source of light that works with white-light viewable holograms. There are a whole range of lights with which to view a white light hologram; some light sources are better then others.

Just remember that white-light holograms require a light source which contains the original exposure wavelength and enough intensity to replay the hologram. Ideal light sources cast sharp shadows, like a spotlight or an average clear light bulb with a single filament. Some very bad light sources, such as fluorescent lights, are extremely diffuse and in some circumstances render white light holograms unviewable.

Whether or not a white light hologram is viewable in bad lighting depends on how the hologram was made. Especially vulnerable are holograms which project an image far out in front of the plate or have great depth. If a hologram like this is illuminated with spotlights coming from multiple sources at different angles, the hologram forms projected images at all the different angles dictated by the spotlights. The mixture of projected images "blurs out" the image the holographer is attempting to recreate.

Hence, holograms made to be viewed in a wide range of lighting, including light from multiple sources, use subjects that have very shallow depth. This is because if there is little or no depth to the object, all the images being created by the different sources appear to be focusing in the same place.

Thus, if you go into a shop that specializes in holograms, one usually finds that the shop has subdued, overhead lighting with spotlights focused on the holograms. This serves the dual purpose of creating a pleasant lighting environment, as well as providing a single spotlight for holograms with considerable projection.

People who display holograms in their homes find an inexpensive way to illuminate them is with a clear light bulb having a single filament. These bulbs are available at any shop with a large selection of light bulbs.

Image Projection of Holograms

Although transmission holograms seem to be more naturally designed to create a hologram with considerable projection, one can make reflection holograms that have a great deal of projection as well. In fact, reflection holograms with considerable projection are a favorite among artistic holographers and the buying public. They are favored because they can be hung on the wall and illuminated just like a painting, whereas transmission holograms need to be lit from behind, often requiring a much larger viewing area.

Laser-viewable transmission holograms demonstrate amazing depth and projection when the correct equipment is used. It should be noted that the depth of the holographic image is not so much a function of the power of the laser as it is the coherence length of laser light (you can read more about coherence length in the laser chapter, Chapter 11).

Theoretically, the maximum image projection in front of the hologram plate can be as great as the projection in back of the plate (depth of the image). Unfortunately, it is difficult for our brains to make sense of greatly projected images. Because of this and the fact that there are some optical distortions in the image planing process, projected distances in transmission holography are usually kept under four feet.

Laser transmission holograms have the widest parallax (the ability to see around an object from side to side) and resolve the greatest depth of objects. There are laser transmission holograms, for example, of people and objects in a 4000 cubic foot room, made by pulsed lasers.

Not surprisingly, projected hologram images like this generate one of the highest shock and thrill responses from viewers. A good percentage of first-time viewers respond by waving their hand back and forth through projected images in disbelief.

Pseudoscopic and Orthoscopic

Both laser viewable and white light transmission holograms share a fascinating property. As you recall, after developing the hologram is put it back in the plateholder. The original object has been removed from the table, yet the viewer can look through the holographic plate and see the object appearing deep within the plate at its original position on the table.

Now comes the interesting feature. Take the transmission hologram out of its plateholder, flip it over, and put it back in the plateholder. Step back and look at the plate. You see the image forming out in front of the plateholder (between you and the plateholder). It focuses in air the same distance in front of the plateholder as it originally sat behind the plateholder. You also see that the image is a *pseudoscopic image*.

What is pseudo scopic? An image as normally seen in everyday life is an orthoscopic image. A pseudo scopic image is the opposite of this. For example, if one's viewpoint moves to the right, you do not see more of the right view of the image but the left view, and when the viewer raises his viewpoint, the lower part of the subject comes into view instead of the upper part.

A pseudo scopic image yields an exciting effect, but it can be confusing to the viewer. Some artists, however, have produced exciting pseudoscopic work using geometric shapes like wide spirals, pyramids and cones.

Recording Materials

Although we have discussed how holograms are made, we have not discussed the material that they are recorded on. In photography, the most common item used to capture images is a silver halide emulsion on a film base. In holography, there are a number of items used to record your image. The most common recording media are:

1) silver halide

2) dichromated gelatin

3) photopolymer

4) photoresist

Note that we use the phrase "recording materials" instead of emulsions. That is because not all the items used to capture holographic images are emulsions. In an embossed hologram, for example, the holographic image is literally stamped into clear plastic. A discussion of the different recording materials requires a chapter of its own, to which we dedicate Chapter 12.

Lasers

We will discuss lasers a little more in depth later in this publication, but it is important that we touch on the subject in the introduction. The type of laser you use affects what subjects you can holograph and we want to discuss that next.

There are two major kinds of lasers; the Continuous Wave (CW) laser and the pulsed laser. The CW laser emits a steady wave of laser light, whereas the pulsed laser emits laser light in bursts.

Continuous Wave Laser: The power of a CW laser is measured in watts (w). The CW laser is by far the most common laser used in holography. In holography labs, most of these lasers fall in the 5 to 50-mw (milliwatt) range. One of the great problems with CW lasers is that they cannot make the extremely short exposures necessary to capture a live subject. Consequently, there must be absolutely no motion at all during the exposure with a CW laser. An exposure with a CW laser can take a fraction of a second to many seconds. Because there cannot be any motion at all during the exposure, we need eliminate any vibration coming from the ground. To do this we make or buy a *vibration isolation table* on which to put our laser, optics, and objects. Since it is absolutely critical that we have no motion at all, the subjects that we holograph with CW lasers have to be "dead" objects. Feathers might move in the breeze and living things simply move too much.

Remember that what we are recording on the plate is the reference beam and the object beam converging (or interfering) with each other at the plate. If the object moves even a microscopic amount (on the order of wavelengths) from one moment to the next, we will record two different interference patterns and the hologram fuzzes out or doesn't even show. The effect is like some photographic daguerreotypes of the 1800's - only much less forgiving.

Pulsed Laser: Pulsed lasers, quite the opposite of CW, emit extremely quick bursts of very powerful laser light. Consequently, the exposure time is much shorter than a CW laser. Exposures can be made in nanoseconds (one nanosecond is one billionth of a second).

You don't need a vibration isolation table for the pulsed laser. What can you shoot? Anything you want. You can shoot an entire room of people belly dancing in costumes of paper with feathers in their hair, and birds flying around the room. Why such freedom? Because your subject can't move significantly in a nanosecond.

What are the drawbacks of pulsed lasers? Why doesn't everyone buy one? The answer is money. They cost about $60,000 and require a lot of extra overhead and care. Lasers don't last forever and when a pulsed laser burns out it is expensive to fix. Holographers are anxiously awaiting, with cash in hand, a low-cost, easily-maintained pulsed laser.

The reason this discussion on lasers is important is that who you go to for making your hologram depends on what you want holographed. If you have a corporate logo or some other object that does not move, a CW laser can do just as good a job as a pulsed laser. There is no need to pay extra to have your hologram made by pulsed laser if you do not need it. On the other hand, if you have a live subject, you must use a pulsed laser.

Choosing a Subject

One last topic we should cover in this introduction is what kind of images you can use to make a holograms.

As with any creative art form there are a whole multitude of choices available, and everything depends on what you want to accomplish. To simplify things, we will list some of the most common items that are used for holographic subjects and then comment on each item. The most common items are:

1) 3-D model (like a sculpture).
2) 2-D model - just like a graphic arts layout for a printing job, complete with overlays.
3) Movies - specially-shot motion pictures for a hologram that moves.

Although some of the above topics are covered in more detail later in the book, we will give an overview of each of them now.

3-D model: This is the most common way to create a subject for a hologram. What your model is, of course, depends on the type of laser you are going to use for your exposure. With a pulsed laser, as we have already mentioned, you can shoot live subjects and just about anything you wish. Most holograms, however, are made with a CW laser and the models used have a dramatically different look depending on the type of paint used to cover the item.

In order to get some idea of what your hologram will look like, many artists paint the subject and then look at it under the light of the laser they are going to use for exposure. Several repaints may be necessary to get exactly what you want. It is possible, with some of the latest techniques, to shoot a hologram of a subject and then reduce the subject in the copying process. Consequently you can have portraits of people reduced and still preserve much of the parallax that holograms have to offer.

2-D model: You will see this method being used with great abundance in embossed holograms. You have the ability to create camera-ready art for an embossed hologram much the same way you would create artwork for a printer, with the different overlays designated to be different colors and at different depths in the final hologram. You can also have a combination of photos, line art and 3-D objects in your final embossed hologram, although the depth of the 3-D object is limited because we are dealing with a thin hologram. Coloration for this process has come a long way in the last few years. Enclosed in this book is an ad featuring a sample of the type of 2D hologram we are discussing. This topic will be discussed in much more depth in the embossed holography chapter.

Movies: Motion picture footage is used in the making of holographic stereograms, which will be covered later on. Suffice it to say that you make holograms of motion picture footage, if it is shot correctly (with the view in mind of using it in a holographic stereogram). Popular usage of this format includes plastic cylinders which rotate, and looking into the plastic you see the image you filmed. It is suspended in mid-air in the center of the cylinder, performing whatever was done in the movie.

It is also possible to make flat stereograms and emboss them, so that as you move the hologram with your hands, the image moves as well. This was done by Sports Illustrated magazine with an embossed hologram sticker of basketball player Michael Jordan on the cover in 1991. As you turned the magazine, Jordan's smile slowly widened.

We have now had a look at some of the basic principals behind making holograms. These principals are applied by practicing holographers in a multitude of ways. Let's go on and examine some of the different types of holograms that are made.

5

Advanced Principles

Looking beyond the basics of holography, there are a number of interesting ways that you can record your hologram. Embossed holography and stereograms are among the most popular.

Embossed Holograms

Probably the most widely-seen example of holograms are embossed holograms. They appear on everything from money to breakfast food boxes and are one of the largest money-makers in the field of holography. Creating an embossed hologram is a very involved process. Due to the popularity of these holograms, we devote Chapter Ten to a discussion of exactly how they are made.

Holographic Stereograms

Creating holographic stereograms is one of the most exciting fields of holography. If you have seen "moving holograms" you most likely have seen one of these. A stereogram is defined as "a diagram or picture representing objects with an impression of solidity or relief." Consequently, a holographic stereogram is a hologram of pictures or diagrams which gives the impression of solidity or relief.

Several techniques are used to make holographic stereograms. Names are given to the various methods and one will hear names like:

- Integram
- Cross hologram
- Multiplex hologram
- Benton stereogram
- Embossed stereogram
- Alcove hologram

As with any trade, one of the problems in holography is understanding the jargon. Sometimes several names are used to describe the same item and there are many special nicknames used to describe special techniques. The names you see above are all holographic stereograms, and the first five are really the same type of hologram. To give a clearer idea about stereograms, following is a brief synopsis of how a Cross holographic stereogram is made.

Making a Holographic Stereogram

Generally, this is the way a 360° Cross holographic stereogram is made:

1) Make a small stage that rotates 360 0.
2) Put the subject on the stage.
3) Set up a regular movie camera on the floor.
4) Film the subject as it turns the full 360 0. Slight motion is possible but the subject cannot make radical or jerky moves. This latter kind of movement creates what are known as time smears in the hologram---places where the subject looks jagged. Slow, even movements when filming yield the best results.

5) Develop the movie film.
6) Make a hologram of the image in each frame of the movie, making sure to shoot at least three frames for each degree of rotation.
7) The holograms will be on a roll of holographic film. Each hologram, as one can see, will be as tall as the width of the roll but of very narrow width.

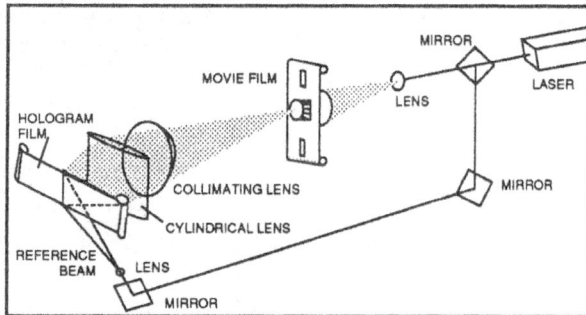

Fig. 5.1 - Optical set-up for a white light holographic stereogram

8) Set up a mask in front of the roll of holographic film to get one narrow hologram (you will be shooting a white light rainbow transmission hologram).
9) Shoot the first exposure, advance the holographic film one frame, advance the movie film one frame, expose again and so forth for the entire movie film.

Viewing a Holographic Stereogram

After development, take the roll of holograms and wrap it around a strong cylinder of clear plastic that is mounted on a display which is able to rotate 360°. Place a clear (unfrosted) light bulb with a single filament in the center just below the holographic film.

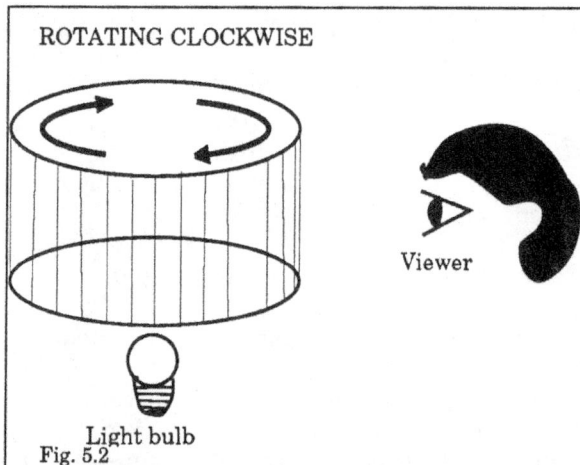

ROTATING CLOCKWISE

Viewer

Light bulb
Fig. 5.2

Turn the light on and rotate the cylinder. Several things are happening here:

• The movie frames are moving past the eyes.

• Each eye sees different images at the same time, thus creating a stereo view.
• Since this is a rainbow white light transmission hologram, the image forms in the same position it was when originally shot (the image usually forms in the center of the cylinder). What we have is a moving holographic stereogram, in which the viewer sees a three-dimensional "movie" of the image moving around in the center of the cylinder.

Other Holographic Stereo grams

Less-Than-360° -Curved-Cross-Holographic Stereogram

The holographic stereograms like the example above can be made in a curved format less than 360°. On the market today are 90°, 120°, 180°, and 360° curved holographic stereograms. The ones that are smaller than 360 0 generally stay fixed and the viewer walks past them to see the motion. The smaller curved stereograms are inexpensive compared to the 360° stereograms.

Flat Holographic Stereograms

Instead of a curved stereogram as described above, one can make a flat holographic stereogram. The procedure for shooting is a little different.

The subject is not on a moving stage this time but on the ground at a distance from the camera. Now we build a straight railroad track on which to put the movie camera. Without aiming the camera directly at the subject, but facing the camera in the subject's direction, the camera moves along the track and takes photos at equal distances. We then develop the film and holograph it much the way we did in the Cross hologram described above.

The result is a flat sheet which, when held up to a light and tilted from side to side, displays an image in motion. With some changes in the holographic process, you may also make this a reflection hologram. The reflection holographic stereogram is viewed by standing under a light, holding the flat sheet and tilting it from side to side. The image can display above the surface of the hologram and move about as it is rocked from side to side.

Embossed Holographic Stereogram

Earlier it was mentioned that rainbow transmission holograms are used as embossed holograms. Since a holographic stereogram is usually a transmission hologram, why not take a flat holographic stereogram and emboss it?

It turns out that this can be done, and it is a popular application. Other techniques along this line include shooting live subjects with a pulsed laser. Very recent developments allow holographers to reduce the size of the image in pulsed shots. In theory one should be able to make a series of live shots, using a pulsed laser, and reduce them to fit an embossed holographic stereogram. Perhaps someday there will be curved embossed holographic stereograms as labels for canned products.

Alcove Holographic Stereograms

What has been done in the stereograms just discussed is to take flat art such as movie film, in which the subject moves from frame to frame, and make a hologram of it. Computers, of course, can make original flat art. Computer-generated images are often easier to work with, particularly when making corrections. Why not generate all the art one needs to make a holographic stereogram using the computer?

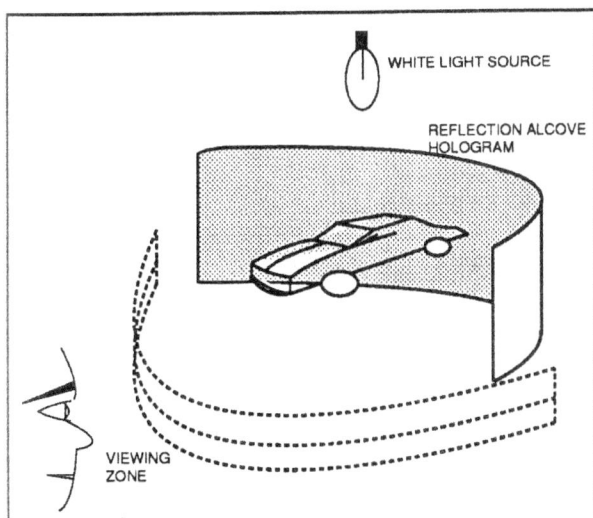

Fig. 5.3- Oblique view of reflection alcove hologram

In a very broad way, this is the concept behind computer-generated holograms (CGH). This is a field of holography that has a great deal of interest at present and we devote a later chapter to discussing the subject in more detail.

One example of CGH is the alcove holographic stereogram. The group led by Steve Benton at the Massachusetts Institute of Technology's Media Laboratory is working hard on this concept and they have already produced some remarkable results. The computer, in this case, creates the flat art. The flat art is then filmed from the video display terminal. The film is then used to make the holographic stereogram. What also sets this hologram apart from the others is the way it is viewed. The image is seen within a clear, concave half-cylinder (an alcove), with nearly 180° of horizontal parallax. In addition to the wide parallax, the image in the hologram can have depth going back through the hologram to infinity. Image distortions are corrected before transfer within the computer.

The alcove hologram is probably the closest holography has come to a totally synthetic, three-dimensional subject to date.

Holodisk®

This is a fascinating application of the reflection holographic stereogram. The process takes 360° of photographic images and holographically integrates them onto a flat Holodisk®. It is lit from above and the subject can appear below, straddling, on, or above the disk surface, within the limitations of the reflection hologram.

A slowly-rotating turntable brings 360° of view to a stationary viewer. An obvious application would be to put the Holodisk~ on phonograph records. As one listens to a favorite song, an image could dance before one's eyes. Unfortunately the revolutions per minute are too high for the Holodisk® to be used on records, but research continues.

Holographic Stereogram Sizes

Sizes vary with each holographic stereogram. The standard curved Cross holographic stereogram is about ten inches high and eighteen inches in diameter with a curved face in 120°, 180° and full 360° formats. The widest film roll one can purchase is a standard 42" wide. Attaching one end of a roll to another, these holograms can theoretically be made in indefinite lengths. Using motion picture footage, one could conceivably run hundreds of feet of continuous holographic film. Current technical restrictions, however, need to be overcome before this can be a reality.

Color Variations in Holograms

An obvious question that most people ask is, how close can holograms come to full-color pictures? The answer is that each of the recording materials are having different degrees of success. Let's look at each of them:

Silver Halide: Full-color holograms in silver halide are not available as far as we know but multicolored silver halides are abundant.

How are multi-colored holograms made? Let us say you want to have two subjects in the hologram. You want each one to be a different color and you want them to be seen by the viewer at the same time (i.e. not multichannel where you have to move the hologram from side to side to get another "channel" and see the other image). For two colors, the exposure would go something like this:

1) The first exposure is made.
2) The subjects are exchanged.
3) The recording material is pre-swelled to another color's percentage.
4) The reference angle is changed and the second exposure is made.

The pre-swelling makes the whole recording surface fatter for the exposure. When the hologram is processed the swelling agent is washed out and the thickness of the hologram returns to normal. At normal thickness, the hologram does not reflect the red wavelengths by which it was made, but instead reflects shorter wavelengths - ranging from an orange-red to violet, depending on how much it was pre-swelled.

For registration purposes the reference angles also must be changed with each pre-swelled color exposure. If you make a two-color hologram by pre-swelling the recording material but do not change the reference angle, the result will be a two-channel hologram. In this kind of hologram, as stated above, only one image can be seen at a time.

There are many complications with the above procedure: swelling agent ratios, changing of reference angles, and multiple exposures. Added to this is the problem that the hologram cannot be copied without going through the whole process again (it cannot be mass-manufactured). Consequently, this method is used for a single hologram or for small-run, custom work. It should be noted that some film suppliers, like Agfa-Gevaert, make pre-swelled emulsions for just this effect.

In sum, full-color holograms, with colors like those seen in a color photograph, are still very difficult to produce in silver halide.

Dichromates: One of the major innovations in the past few years, according to a leading manufacturer of dichromate holograms, is full-color capability.

Photopolymers: There has been some progress in photopolymers with color holography and we have seen multicolored photopolymer holograms. DuPont claims in their literature to have made "significant improvement" in full color film. Samples of the film are being offered for evaluation.

Embossed holography: There has been good success in producing results close to full-color pictures in embossed holography. There have been numerous full color stereograms produced. One of the first was a flat color stereogram "The Clown" by Sharon McCormack) where the figure is both in color and moves as you move the hologram from side to side.

SECTION 3

Holographic Applications

6

Holographic Optical Elements

One of the most financially rewarding fields of holography has been the making of Holographic Optical Elements, commonly referred to as HOEs.

Often a hologram can mimic what conventional lenses, mirrors and other optics do more efficiently and at a lesser cost. In this chapter we explore some of these options.

As the name implies, HOEs are optical elements such as lenses, mirrors, etc. that are made holographically. Although the fabrication can be quite involved technically, the concept of the HOE is relatively simple. Following is a simplified explanation.

HOEs: The Concept

We know from our earlier discussion that in making a transmission hologram the light from the object beam reflects off our object at innumerable points, and goes on to strike the photosensitive plate at the same time that the reference beam does. We then develop the plate to get our hologram. Whenever we shine a reference beam on our hologram at the same angle used to create the hologram, the light from the reference beam focuses in space to visibly recreate our original object.

We can say, then, that there are innumerable points of light coming from the object which together form the object beam. After the hologram is developed and we illuminate it, the transmission hologram refocuses the reference beam in such a way that it recreates the points of light that were reflected off the object. What we actually see is a compilation of countless points of light, each with its own focal length, being refocused in space by the hologram.

Let's consider a much simpler transmission hologram. Instead of an object which reflects many points of light, suppose our object has just one point of light. If you stop to think about it, the transmission hologram of a single point source of light functions in the same way as a traditional concave lens (illustrated in fig. 6.1).

You will also find that a reflection hologram functions in much the same way as a convex mirror, as shown in fig. 6.2.

These holograms are HOEs. They are generally made without objects and with the intention of mimicking a conventional optical lens or a variation of conventional optics. With an HOE, the method for producing the image differs from conventional optics, but it must conform to the same basic principles. These include such fundamentals as the equation for determining the focal length of the lens.

As you can imagine, there are differences between conventional lenses and HOEs. Here are a few of them:

• The angle of light used by a holographic lens is very selective, since it is the reference angle.

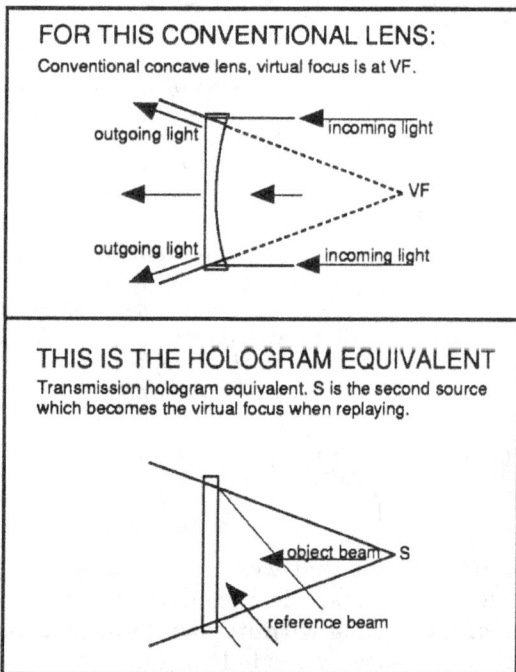

FOR THIS CONVENTIONAL LENS:
Conventional concave lens, virtual focus is at VF.

outgoing light   incoming light
VF
outgoing light   incoming light

THIS IS THE HOLOGRAM EQUIVALENT
Transmission hologram equivalent. S is the second source which becomes the virtual focus when replaying.

object beam S
reference beam

6.1 Conventional concave lens, and holographic equivalent.

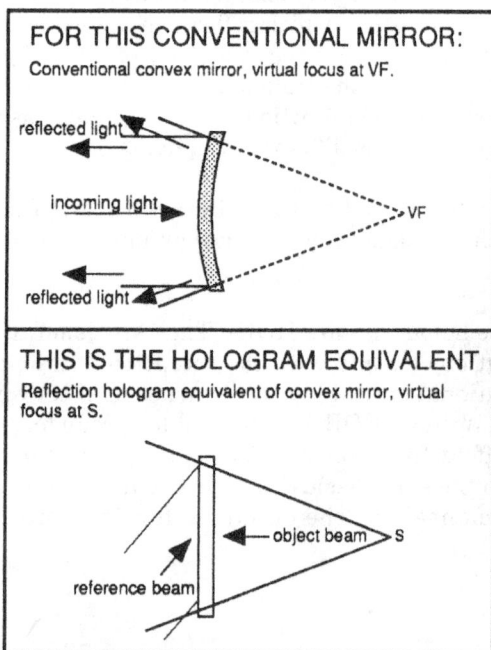

FOR THIS CONVENTIONAL MIRROR:
Conventional convex mirror, virtual focus at VF.

reflected light
incoming light   VF
reflected light

THIS IS THE HOLOGRAM EQUIVALENT
Reflection hologram equivalent of convex mirror, virtual focus at S.

object beam S
reference beam

Conventional mirror and holographic equivalent.
Fig. 6.2

• The frequency of light used by a holographic lens can be very selective since it is the frequency used to make the hologram.

• Unlike a conventional mirror, a holographic mirror can be rotated 180 degrees to form a real image with a positive focal length.

Advantages of HOEs:

• Since you can create an HOE lens that, unlike conventional lenses, is very angle-and frequency-specific it makes possible the construction of unique and very selective optical system configurations. One example is an HOE mirror that reflects light coming in from an angle greater than 45 degrees, or a mirror that reflects a given percentage of a pre-designated, specific wavelength.

• Unlike conventional optical elements, the function of HOEs is relatively independent of the substrate geometry. Consequently, HOEs can be produced on thin substrates that are relatively light even for large apertures.

• Since HOEs are holograms, spatially overlapping elements are possible because several holograms can be recorded in the same layer.

• HOEs can correct system aberrations, so that separate corrector elements are not required.

• HOEs offer the possibility of mass production at a significantly cheaper unit cost than conventional optics.

• Some HOEs can be produced from computer-generated holograms. When the waveforms needed to form the HOE are calculated and generated by a computer, an HOE that has little noise and imperfection can be produced.

In order to make an HOE that reproduces a desired optical setup, there are two approaches to take. One is to physically set up the optical path that you wish to reproduce, using conventional optics, and make a hologram of it. In this case, the setup itself serves as a master for a hologram which can be mass-produced for a far lower unit cost. For many holograms, this is still the easiest and most accurate method. The disadvantage of this method is that all of the optical components still need to be purchased for the master setup.

The other way to make a wide variety of optical components holographically is to reproduce the optical properties desired and follow the same optical laws. There are many different types of HOEs. In general they are either gratings, lenses, mirrors, beam splitters or combiners, but almost anything is possible.

Diffraction Gratings Diffraction gratings are among the simplest HOEs to make. For a very simple diffraction grating, you only need the interference of two or more beams of laser light, such as in the general setup below.

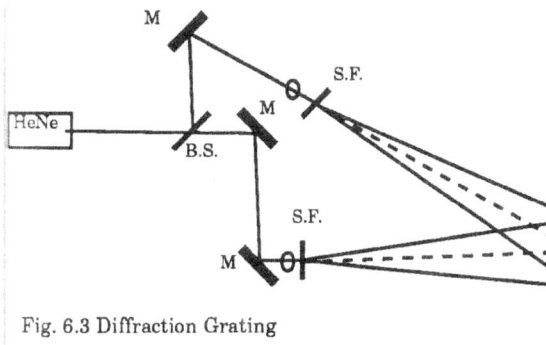

Fig. 6.3 Diffraction Grating

A diffraction grating is the holographic equivalent of a prism. If a diffraction grating is shot with a HeNe laser, when it is reconstructed with the same laser at the same angle, the light will diffract at that original angle. If, on the other hand, it is reconstructed with white light, only the red light with which it was made will diffract at the original angle. The shorter wavelengths of light will diffract progressively less until we are out of the visible spectrum (see fig. 6.4).

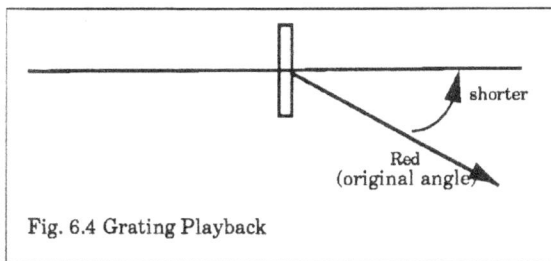

Fig. 6.4 Grating Playback

Holographic Lenses

As mentioned earlier, a transmission hologram can be much like a conventional lens in form and function. It can be mounted in much the same way and placed as a conventional lens, yet cover a far greater area at a much lower cost. A holographic lens is called a Gabor Zone (see fig. 6.5). There are two simple techniques for constructing holographic lenses, both of which derive from the lens equation:

$$1/F = 1/U + 1/V$$

where F is the desired focal length, U is the object distance and V is the image distance.

The first technique is to use a collimated beam for one of the beams. Since $1/00 = 0$, therefore F=V. In other words, the focal length of the lens is the distance of the spatial filter from the holographic plate. (see Fig. 6.5).

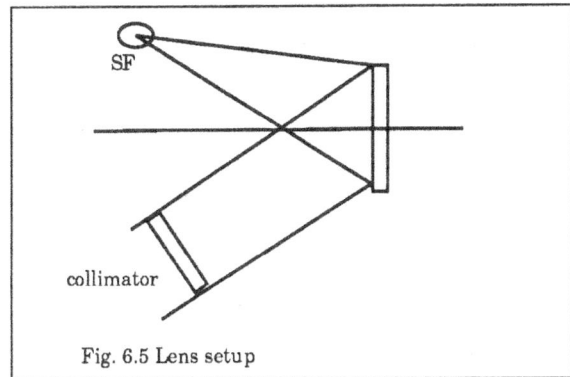

Fig. 6.5 Lens setup

A more general setup is to use two finite sources at different distances from the plate such that:

$$1/F = 1/D1 - 1/D2$$

The reason for the subtraction here is that, on reconstruction, the reference beam distance is reversed. An algebraic trick here is to use D2 = 2Dl, so that:

$$1/F = 1/D1 - 1/2D1 = 1/2D1 = 1/D2$$

i.e. F = D2, but this is not essential. (See Fig. 6.6)

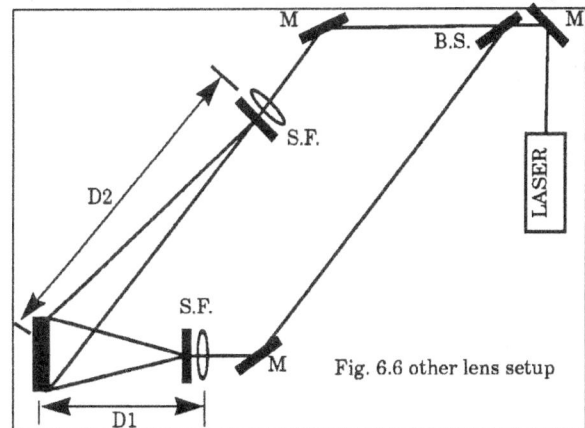

Fig. 6.6 other lens setup

Holographic Mirrors

Both the above formula and its concept are also true of holographic mirrors. One only has to convert the setup to a reflection hologram setup, by illuminating the back of the plate with the reference beam to produce mirrors.

Again, if D2 = 2D1, then the focal length is going to be equal to D2. You can also keep in mind that using different powered microscope objectives at appropriate distances can also create the desired effect. (Fig 6.7)

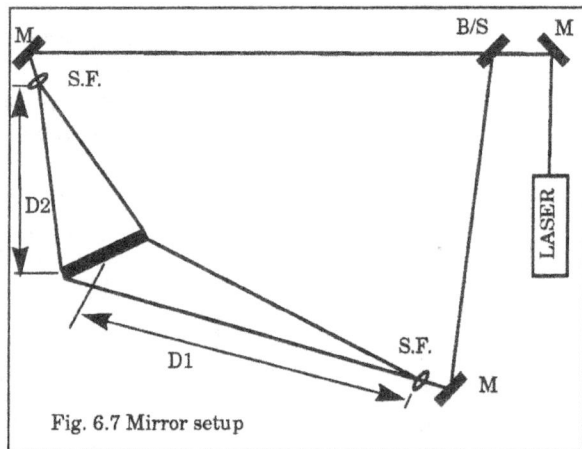

Fig. 6.7 Mirror setup

Holographic Collimators

It is worthwhile for us to notice at this point that a collimated beam has the property D = 00. What can be derived from this is that it is possible to approximate a collimated beam holographically, by moving one of the beams very far away in relation to the other beams. At this distance, the spherical waves coming from the spatial filter are almost parallel and approximate a plane wave front, or collimated beam. This kind of geometric approximation is often enough for a given application. This can be done with either the lens setup or the mirror setup, but since the distance is great and therefore the power low, the higher efficiency of the mirror collimator is recommended.

Holographic Mirror Beamsplitter

Another essential component of a two-beam holographic setup is the beam splitter. Usually this is a half-silver or a specially-coated mirror, whose functions can be mimicked holographically. The procedure is to make an HOE mirror with beams that enter the plate perpendicular to one another. This can be done using only a single beam in this simple configuration, a useful trick since, without a beam splitter, the beginning holographer has only one beam.

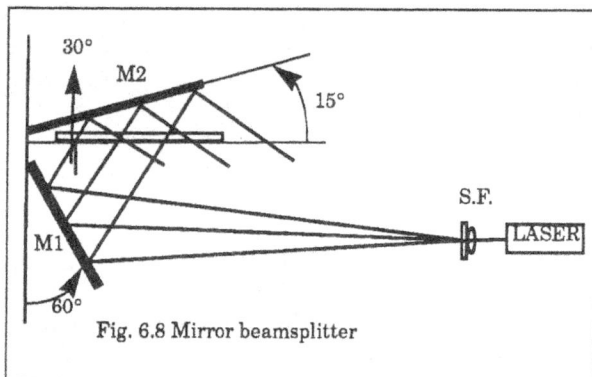

Fig. 6.8 Mirror beamsplitter

In Fig. 6.8, Mirror 1 is a front surface mirror at 60° to the holographic plane. It is far enough away from the source that a collimated beam is approximated. This means that the waves passing through the holographic plate are plane waves. Mirror 2 is at -15° to the holographic plate, so that the incident rays at the plane are perpendicular to one another.

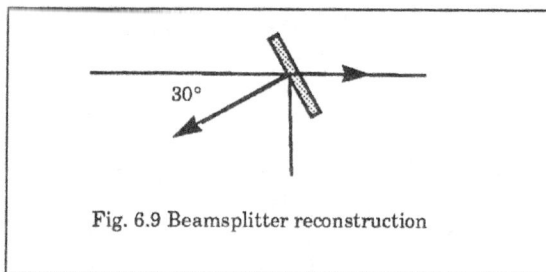

Fig. 6.9 Beamsplitter reconstruction

As you can see in fig. 6.9, when reconstructed at the shooting angle the HOE reflects a portion of the beam at 90° to the incident.

A variable beam splitter is often useful and can be made by moving the source off-axis, so that the strength of the beam across the holographic plane reduces as it gets further away from the source. Then you have a beam splitter that will transmit and reflect more or less of an incoming beam as it is moved along it own axis.

The beamsplitter provides an excellent example of the cost-effectiveness of HOEs. In this case, what you would have to purchase (other than your table and laser) are two front surface mirrors of a few square inches each (approximate cost $20). These mirrors are re-usable in many other applications, so the total cost of our beam splitter (without a mount) is negligible. Beamsplitters from large supply houses cost between $500 and $2000 and, of course, can be broken or damaged.

Other HOE Applications

HOEs have many applications other than just simple table components in our modern marketplace. Among others, there are automobile and defense applications, artistic applications and indus- trial applications. We will present a few examples.

Tandem Optics: In many situations it is desirable to be able to look at a volume of material interferometrically. One way to do this is to make a set of gratings that diffract light at precise, controlled angles and efficiencies and position them in such a setup as fig. 6.10.

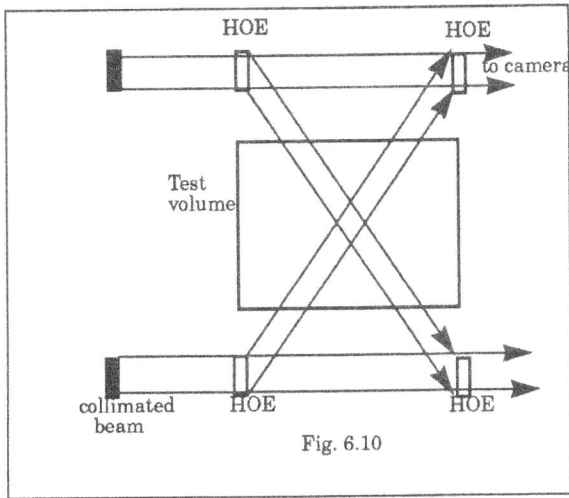

Fig. 6.10

The purpose of this is to establish a dual object beam passing through the material and a dual control (reference beam) to provide the interference.

Another application of the same procedure is to provide what is called the Perfect Shuffle optical system, whereby irrelevant information can be eliminated and relevant data stored in the correct place by a series of HOE diffraction gratings that shunt the light at different angles in the system (Fig. 6.11).

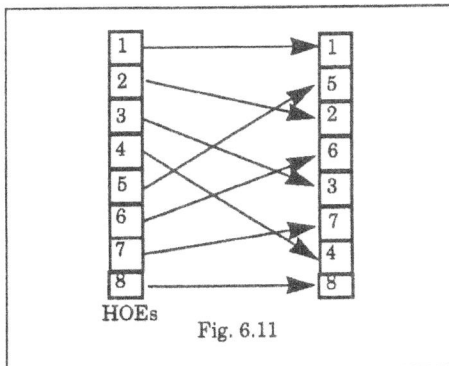

Fig. 6.11

Artistic applications: One of the problems in holographic artwork is that the holograms with the best depth, silver halide laser viewable holograms, are also the hardest to display. This because of the amount of space needed behind them. Again, a grating is a possible solution; in this case, a grating designed to receive an incoming beam at a very shallow angle and diffract it to the reference angle of the art hologram. This HOE can be sandwiched to the original hologram and sold as a package (fig. 6.12).

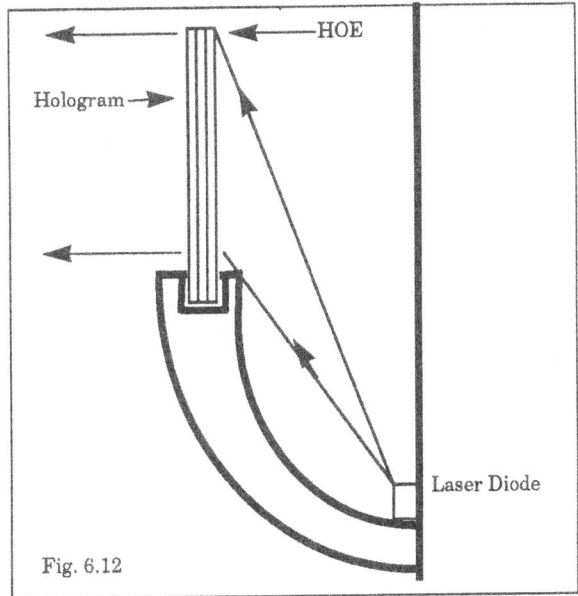

Fig. 6.12

Military and Automobile Applications: Head-Up Displays (HUDs) have been common in aircraft for quite a few years, and they may work their way into high-end automobiles.

The idea is that a CRT display can be projected onto a glass substrate at eye level, which will neither require the driver to take his eyes off the road nor interfere with his view of the road. Following this chapter is one devoted specifically to HUDs.

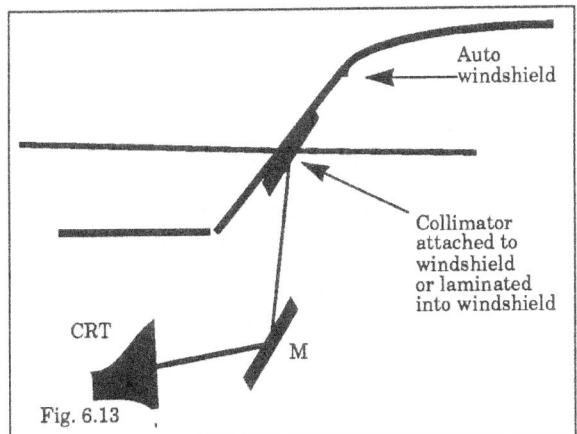

Fig. 6.13

This chapter was contributed by:

Phil Burfoot

El Cerrito, California, USA

References

Saxby G. Practical Holography. (1988) Prentice Hall International. 228-234.

Stojanhoff, e.G., & Windeln, W., Development of high efficiency holographic optics. SPIE Proceedings # 812 (1987)

Kobolla, Harald., Sauer, Frank., & Volkel, Reinhard. Holographic Tandem Arrays. SPIE Proceedings # 1136 (1989)

Jeong, Tung & Wesly, Edward. HOE for Holography. SPIE Proceedings # 1183 (1989)

Del Vo, P., Rizzi, M.L., & Stibelli, S. Holographic Head-up Display for the Automotive Industry. SPIE Proceedings#1136 (1989).

7

Head-Up Display (HUD)

A Head-Up Display (HUD) allows the viewer to see display information superimposed on an outside scene. The display image is located at a distance so that the viewer can keep his eye focused on the distant scene while he views the display.

How it works

A Head-Up Display projects the display image onto a partially transparent screen called a combiner that reflects the display to the viewer while allowing the viewer to see through to the outside world. Fig A shows the elements of a typical aircraft HUD.

The advantage of a HUD is that it allows the -:ewer to see the projected display information while still looking at the scene beyond. An example of how useful this can be is in allowing a pilot to see both the runway and his instruments simultaneously during landings.

Another advantage is that the distant display image saves the time needed to refocus the eyes between nearby instruments and the world outside. .Some years ago commercial pilots voted the HUD as the most important addition that they would like to have to their cockpit instrumentation.

History

Head-Up Displays were first used in fighter aircraft to allow the pilot to fly the plane and operate the weapon systems simultaneously. Early HUDs used a partially reflecting conventional mirror as a combiner. Later HUDs were improved by using a reflection hologram as the combiner.

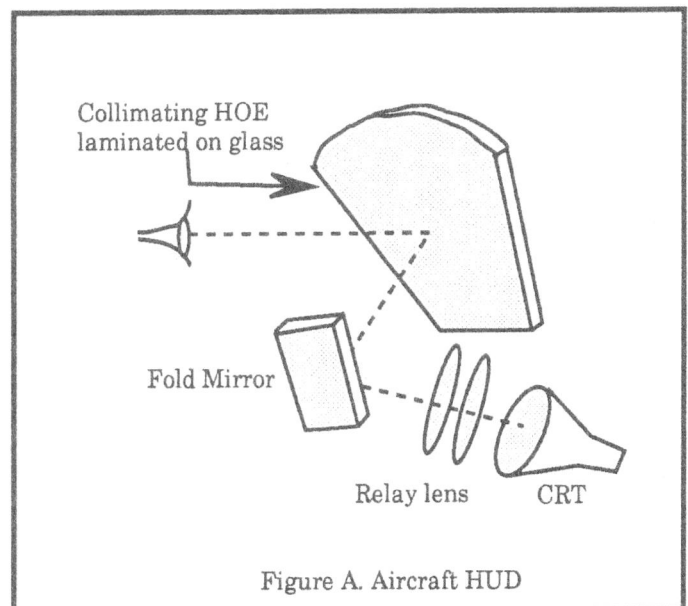

Figure A. Aircraft HUD

(labels: Collimating HOE laminated on glass; Fold Mirror; Relay lens; CRT)

The holographic combiner for HUDs was pioneered by Hughes Aircraft in the 1970's for a Swedish fighter program. The holographic combiner gave the pilot a wider field of view, increased brightness and improved seethrough than did previous conventional mirror HUDs. The reflecting hologram in front of the pilot acted as an aspheric optical element that provided the accurate display information needed to aim weapons. It also could display forward-Iooking-infra-red (FLIR) information with a wide enough field-of-view to allow high-speed, low-level night flight. The increased workload in single seat aircraft made the improved performance offered by the holographic HUD a necessity.

Holographic advantages

One of the advantages of a holographic HUD combiner is its ability to reflect only a very narrow wavelength spectrum. This means that the reflectivity can be very high for the wavelength of the display while still remaining very low for all other wavelengths in the field of view.

By using a narrow band display source such as a P43 phosphor cathode ray tube the HUD display can be both very bright and very transparent, with minimum coloration of the see-through scene. In an aircraft, the improved brightness makes it easier for the pilot to see his display against a bright background such as a sunlit cloud. The improved see-through gives him excellent forward vision.

In addition to its wavelength selectivity advantages, the hologram can provide optical power so that it can perform as a collimating mirror close to the pilot's eye. This would show a wider field of view than previous HUDs, in which conventional collimating lenses had to be located further away in the body of the aircraft.

Related Displays

There have also been visor displays built for the U.S. Air Force and Navy in which a holographic combiner is fabricated on a visor with a miniature cathode ray tube (CRT) mounted on a helmet to provide what is effectively a HUD on a helmet (Fig B).

In another version built for the U.S. Army, the CRT was replaced with a night vision image intensifier tube to provide a night vision visor display in which the wearer sees an intensified image superimposed on a transparent holographic visor (Fig C).

Another type of display that uses a hologram is what might be termed a static HUD. In this application, a hologram containing an image or text is embedded in a transparent windshield. This image can be "turned on" by illuminating it with a light on the instrument pad (Fig D). This simple display, if put in a car, can display data, such as turn signal flasher arrows, or warnings, such as "low fuel" or "engine overheat". One problem with this type of display is preventing spurious turn-on from ambient sources such as sunlight.

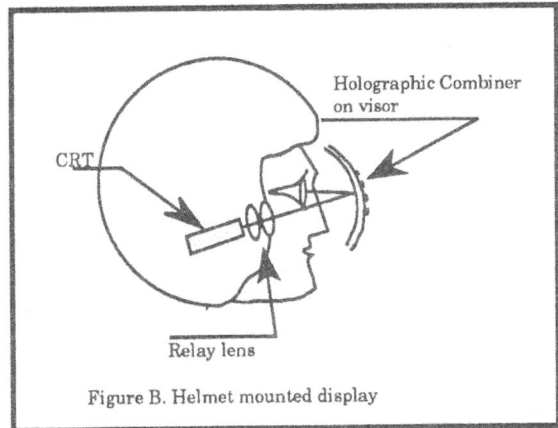

Figure B. Helmet mounted display

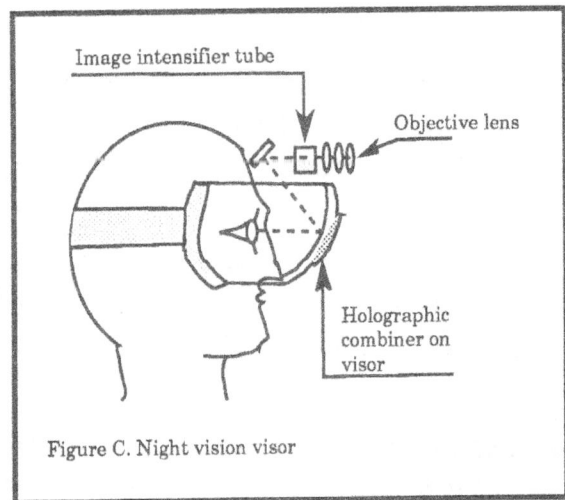

Figure C. Night vision visor

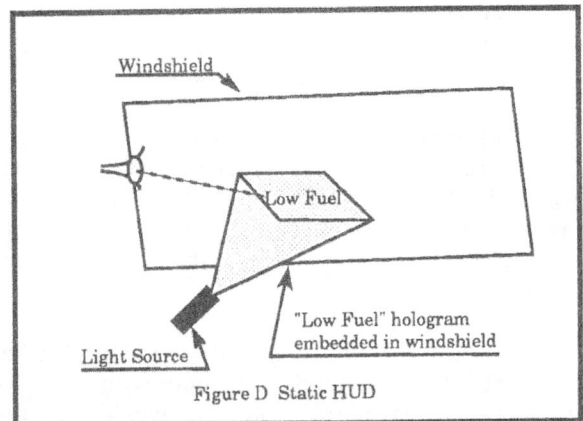

Figure D Static HUD

CURRENT STATUS OF HUDS

Military

The most significant use for holographic combiners is still for military applications such as fighter aircraft HUDs or tank displays. A number of companies have made military holographic HUDs including Hughes Aircraft and Kaiser (USA), GEC and Pilkington (England) and Thomsen CSF (France).

Commercial

Flight Dynamics in Oregon has made holographic aircraft HUDs for commercial aircraft and X-IAL in France has made experimental HUDs for high speed European trains. More on Flight Dynamics' research is included at the end of the chapter.

Aside from aircraft, the greatest interest in comercial holographic HUDs is for automotive applications. At the last (1991) Tokyo Auto show, ten different manufacturers showed experimental or production HUDs, of which seven used holographic elements. Most of the holograms were used as combiner elements on the windshield, although a production Toyota HUD in the (Japan only) Crown Majesta used a hologram as a curved selective mirror in the optical system. In previous years European manufacturers such as Volkswagen and Citroen have shown experimental HUDs with holographic combiners. A typical automobile HUD is shown in Fig. E.

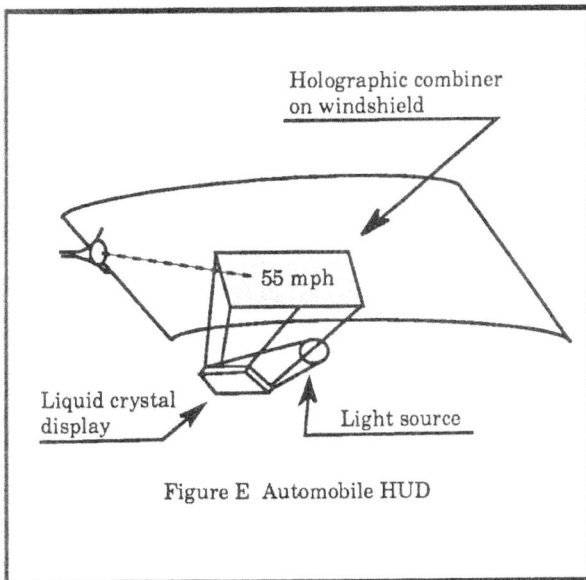

Figure E Automobile HUD

In spite of all this activity, the only three automobile HUDs that are in production do not use holographic combiners. Nissan and oyota use partially reflective sol-gel coatings and General Motors uses the reflection of an untreated windshield.

Materials

Since good visibility through the HUD with minimal scatter is important, combiner holographic optical elements (HOEs) must be volume rather than thin surface holograms. This means that low-cost embossed holo- grams are not acceptable. The more complex process of optical recording is needed for quality holographic combiners. The recording material which gives the highest efficiency with the best see-through is dichromated gel- atin (DCa). This is used for all aircraft HOEs. DCa is very durable when sealed between glass plates. Some military HUD combiners that are nearly twenty years old still perform well in their original test aircraft.

The next-best recording materials are the photopolymers such as DuPont's Omnidex, or Polaroid's DMP 128 which, although of lower optical performance than DCa, have simpler post-exposure processing and better resistance to moisture.

Experimental automotive HUDs have been made in both DCa and photopolymer. The automobile manufacturers all want a HUD to be part of the windshield instead of the stand-alone combiner usual on aircraft. This means that the holographic material must withstand the heat and pressure in the production lamina- tion of automotive windshields. Alternatively, a holographic combiner could be added as a decal to the surface of the windshield if it could withstand the abra- sion, humidity and other weathering factors.

WHY AREN'T THERE MORE HUDS IN PRODUCTION?

Since holograms can make superior combiners for HUDs, and are used in military aircraft, why aren't they used in any production automobile HUDs?

Cost

First, making a high-quality volume hologram is more expensive than making a simple reflective coat- ing. Second, even though the performance requirements are less stringent, the hologram fabrication problem is, in some ways, worse for automobiles than for aircraft. This is because the aircraft combiner is a small stand- alone plate while the car combiner is part of the large windshield. The aircraft combiner can be coated with DCa and the hologram recorded in place. This is not practical for a large windshield. Government regula- tions for various standard safety tests also limit the fab- rication processes that can be used to install a hologram in an automobile windshield. The most inexpensive manufacturing method for a car is probably to make holograms on plastic film, which would then be laminated inside the windshield. Problems are: stretching of the film in the curved laminate, heat and pressure during lamination and compatibility with the windshield laminating material.

Durability

Holographic recording materials are generally less durable than many inorganic reflective coatings such as those used by Nissan and Toyota in their HUDs. Even when sealed in a windshield, solar radiation can degrade some holographic materials.

Optical Performance

Although a HOE can provide better combiner performance, there are a number of potential optical problems associated with using a holographic element on a windshield that must be carefully addressed.

• In an aircraft, the HUD combiner is a separate, fairly small element which can be tilted and shaped for optimum holographic performance. In a car, the holo- gram must be put on a large windshield which was shaped for styling and aerodynamics rather than holo- gram system optical performance. This means that the hologram's recorded fringe pattern may have to depart significantly from the windshield surface shape, which aggravates various problems such as unwanted diffraction.

• Transmitted light from the sun or other outside sources such as oncoming headlights can be diffracted by the hologram into a rainbow flare.

• The see-through image can be badly colored by the removal of too much transmitted light if the hologram reflects too broad a spectrum.

• The photopic see-through can be reduced too much.

• If a wide spectral-band source is used, such as a low-cost vacuum-fluorescent display, the dispersion of the various wavelengths can cause a fuzzy image.

• The narrow band hologram is inherently better at showing a monochromatic image, while a multicolored image may be desired for car HUDs.

• When viewed from outside the car, the holographic element can reflect a colored patch on the windshield which might be considered unsightly.

Lack ofFunctional Design development

The current applications for automobile HUDs use them simply as a redundant display instrument, showing speed, "flasher-on" indicators, or warning signals that are already indicated elsewhere. To make a holographic combiner worthwhile, the HUD needs to have

a greater value which requires rethinking the display and control requirements to make more use of the HUD capability. Some legislation changes may be needed to allow a HUD to become the primary display instrument.

THE FUTURE OF HUD

What is the future for holograms in Head-Up Displays? Dividing the market into high-cost and low-cost applications yields different answers.

High Cost

Holographic combiners will continue to be used for military HUDs because they provide the best brightness and see-through performance. The production numbers are small compared to potential commercial applications, and the total system cost is high enough so that high-quality dichromated gelatin holograms can be virtually hand made for these combiners. Commercial aircraft HUDs and even locomotive HUDs cost enough so that they can probably be successfully produced with the same DCa technology that has been developed for mili- tary use. The problem is that with military cutbacks there may be no foreseeable future market for fighter aircraft HUDs. Further, with financially- strapped airlines laying off pilots and grounding aircraft, there is no near-term HUD market for com- mercial aircraft either.

Low Cost

Developing low-cost holographic HUDs suitable for the commercial market such as automobiles is very difficult. Perhaps the first necessary step is for the car manufacturers to better identify what the car HUD can do for the driver, so that the driver wants and needs a HUD. The key may be to use the HUD not just as an advisory instrument, but as part of the driving system to allow the driver to con- trol all the car's functions without taking his eyes off the road.

The need for a HUD is here. The modern car, with all its entertainment, communication and other systems such as: AM-FM radio, tape deck, disc player, cellular telephone and air conditioning is getting more complicated all the time.

And more complexity is coming. In Japan several satellite-linked systems which present navigation map displays to the automobile driver are already in production. Conventional head-down controls for all these functions are becoming dangerously distracting for a driver. As a solution, selected, head-up information, perhaps coupled with tactile or audio controls, could be presented that would allow total head-up control of an automobile. The HUD could then even become a mandatory safety device by allowing the driver to keep his attention on the road ahead.

Once the HUD is established as a valuable and needed part of an automobile, then the extra cost of a holographic combiner to improve the performance of the HUD may be justified. Trade-off's then need to be made between various technical performance issues and the recording material and method for low cost manufacture and installation of the holograms in or on the windshield. The results of these technical and production issues will determine whether future commercial HUDs use holographic combiners or merely use a partially reflective coating. If solutions can be found, then commercial HUD combiners should be a growing market for holographic optical elements.

This article was contributed by

Gaylord E. Moss
PO Box 9130
Marina Del Rey, CA 90295 USA
Tel (310) 827-3983
FAX (310) 827-3983

Mr. Moss worked for over two decades in developing HUD applications for Hughes Aircraft Company. He now works as a consultant on holographic applications.

HOE Applications: A Case History

Flight Dynamics, Portland, Oregon

HUDs were first introduced as military aircraft displays for the purpose of projecting the image of a target reticule out in front of the aircraft. These types of HUDs consist of an illuminated fixed retizule (later replaced by a CRT), a collimating lens, and a flat glass combiner. The purpose of the combiner is to reflect some of the light from the reticule while transmitting a large amount of the light from the outside scene. In order for the combiner to be transparent, the amount of reflected reticule light is usually only 20%. When CRTs were introduced, the low reflectivity required driving the phosphors at maximum brightness and replacing them frequently.

Flight Dynamics, which manufactures HUDs for the purpose of guiding commercial airliners in low-vsibility conditions, is one of a very few companies using holographic coatings on HUD combiners. The

reflection hologram, due to its high efficiency and narrow spectral reflection band, can be matched to the spectral outlet of a CRT phosphor to allow greater than 50% reflection of the display light, while maintaining greater than 80% transmission of light from outside the airplane.

In addition, the hologram can be placed on a curved combiner which then becomes part of the collimating optics. This allows for greater field of view. However, the aberrations can be so high that the hologram might be required to have diffractive power independent of the focal power of the curved combiner on which it has been placed. This diffractive power is analogous to using an aspheric reflector, but is simpler to fabricate than an a sphere.

Another application of reflection holograms in an HOE is called a spectral notch filter. This type of HOE is used primarily for filtering out bright laser noise from broadband signals. Examples would be laser fluo- rescence spectroscopy, eye protection, astronomy, and laser radar. Holographic spectral notch filters are unique in that their efficiencies are very high (typically transmitting less than one part in 10,000 at the design wavelength) and their reflection bands are narrow (around 20 nanometers in the visible).

Some Disadvantages of HOEs

1. Transmission HOEs typically have low efficiency compared to lenses, so applications requiring low light loss benefit from using lenses.

2. In a lot of cases, it is simpler and cheaper to buy a lens than to make an HOE, mainly when the equivalent lens is very simple.

3. While spectral selectivity might be an advantage for some, it is a big problem for the many applications which utilize a band of wavelengths. Achromatic HOEs are manufacturable, but not cheaply or easily.

Contributed by
Robert Brown
Optical Engineer
Flight Dynamics
16600 SW 72nd Ave.
Portland, OR 97224 USA
Tel (503) 684 5384

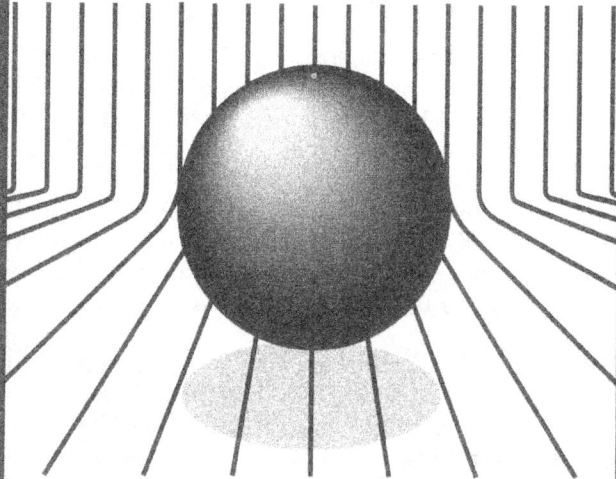

8

Computer-Generated Holograms

Since a hologram is merely a recreation of light patterns, why not plot these patterns on a computer screen and avoid the complicated optical set-up? That's the idea behind computer-generated holograms, which are frequently called CGH.

Thick and Thin

As we explained in the introduction, some types of holograms sacrifice depth so they may be seen in a variety of commonly-encountered lighting conditions. For the sake of convenience, authors sometimes divide holograms into two broad categories: "thin"holograms and "volume" holograms.[1] Basically, holograms are considered thin or 2D when the thickness of the recording material is small in relation to the average spacing of the interference fringes (less than 1:1). Volume holograms have a ratio greater than 1:1.

Thin holograms have little or no depth. Volume holograms can have great depth. Although volume holograms have all the glamorous 3D qualities, thin holograms have their own reasons for being popular. Thin holograms are:

• Easy to view, even under harsh lighting conditions.

• Easy to make, since we are only dealing with 2D subjects.

• Marketable holograms, because many are HOEs like gratings, mirrors, and lenses.

Let us consider only the thin holograms, since CGH are thin holograms. With a fairly thin hologram, we can look at the emulsion and think of it as a flat, two-dimensional object like a piece of paper. We can store all the information for this flat hologram in a computer by simply recording, on an (x, y) coordinate graph, the size and location of all the holes in the emulsion.

Following is an overview of how a computer-generated hologram (CGH) is made.

1) Input Hologram

We either have an existing physical object, or an imaginary, digitized object.

Real Object: If we have a real object, we make a "thin" hologram at this point. Then we have the holographic patterns on the plate digitized by scanning the plate using a microdensitometer. A microdensitometer measures the percentage of light that goes through a material in an extremely small area.

Imaginary Object: If we want to make a CGH of an imaginary object, we calculate the wave patterns at the plate surface by using theories appropriate for Fraunhofer, Fresnel, or near-field diffraction. It should be pointed out that the more complex our object is, the more difficult this task will be.

2) Edit Hologram

If we need to edit or change our hologram in any way, we do so in the computer at this point.

3) Output Computer Generated Hologram

We now output or reproduce our hologram onto some media. Eventually we want to wind up with a transparency that has our holographic pattern on it. There are several techniques used to map the hologram pattern which are covered later in this section. The most common hardware used to output CGHs are:

- Computer-driven plotter
- Computer-driven laser beam
- Computer-driven electron beam
- Computer monitor: displays hologram; photograph is made from the monitor

4) Reconstruct and Detect Hologram

Once the hologram is output, we copy the hologram and mass-produce it, or illuminate the hologram and record it using means that suit our purpose. Detectors may include:

- Eye-simply view the hologram if that is the purpose
- Charged-couple devices
- Photographic film
- Small antennae and receivers have been used for detecting microwaves

Advantages: The advantages of the computer-generated hologram are clear. Here are six main reasons:

1) Objects: No objects are necessary in the simple cases where we can calculate the wave patterns by computer.

2) HOEs: Since we don't need to use real objects, idealized waveforms can be made. This eliminates unwanted extraneous light and "noise" one would get from a normal hologram. This is of great value if we are making HOEs (see also Chapter 5, HOEs).

3) Editing: We can digitally edit our hologram in the computer.

4) Storage: We store our holograms electronically instead of on time sensitive emulsions.

5) Transmittal: We can send our holograms electronically worldwide in a matter of moments.

6) Translation: As we look into a Fourier Transform hologram we see a unique feature. We see a central spot of light and two images, one on each side of the spot. One object is an inverted and pseudo-scopic image of the other. If we rotate the plate clockwise, the images rotate with the plate. If we simply turn the plate so back becomes front and vice versa, like we turn the pages of a book, the images stay fixed in space. This unique feature is used in some specialized cases.

Disadvantages: Here are two major disadvantages of computer-generated holograms:

1) Hologram thickness: The major disadvantage to computer-generated holograms is that they currently only work for thin holograms. One of the great attractions to holography is the "volume" or three dimensional qualities of a hologram. It is true, however, that there are clever ways to create depth using thin holograms such as stereograms. One of the most notable current items is the computer-generated alcove hologram being done by Dr. Stephen Benton at MIT's Media Lab.

2) Computer Time And Storage: Generating the holographic fringe structure of a simple object such as a toy car can require tremendous amounts of computer memory and time, making it difficult and not cost-effective for many applications.

Techniques for Outputting Computer-Generated Holograms

As mentioned earlier, there are several techniques for plotting or mapping the holographic pattern we send to our output hardware. Four of the most commonly-used methods are:

- Binary Detour Phase Holograms
- Kinoform
- ROACH (Referenceless On-Axis Complex Hologram)
- Computer-Generated Interferograms

Binary Detour Phase Holograms

This method was first reported in 1966. The final product is an opaque mask with transparent holes or apertures that represent the hologram.

To output this, the computer calculates the image of the hologram mathematically using a Fourier transform. The paper or media on which the computer prints the hologram is subdivided into miniature cells. In each cell, the computer prints a dot which later becomes an aperture. The magnitude of the Fourier transform at the center of the cell is calculated and that calculation determines the height and width of the aperture. The lateral position of the aperture within each cell is proportional to the transform's phase at the center of the cell.

On photoreduction, the black dot becomes a clear aperture on a black background. Representing the phase by using a lateral shift of the aperture within the cell led to the name "Detour Phase Hologram", an analogy to diffraction gratings with unequally spaced rulings. Another term for these holograms is "binary hologram" because any point on the hologram we create has a transmittance value of zero or unity. There are several variations of the Detour Phase Hologram which allow for a more refined output.

Advantage: It is possible to use a simple pen-and-ink plotter to prepare the binary master and problems of linearity do not arise in the photographic reduction process.

Disadvantage: This method is very wasteful of plotter resolution, since the number of addressable plotter points in each cell must be large to minimize the noise.

The Kinoform
If the object is diffusely illuminated, the magnitudes of the Fourier coefficients are relatively unimportant, and the object can be reconstructed using only the value of their phases. This gave rise to the Kinoform. To record a Kinoform, the computed values of the phase are recorded on a multilevel gray scale, which in turn is used to control a photographic plotter that exposes a piece of film. The master is then photographed again, to reduce it to final size, and bleached to convert the gray levels to corresponding changes in optical thickness.

Advantage: Kinoforms can diffract all the incident light into the final image.

Disadvantage: Less information is available since only the values of the phases are used. Also, if there is any error in the recorded phase shift, light is diffracted into the zero order which can spoil the image by creating an extremely bright light in the center of the hologram.

ROACH
Referenceless On-Axis Complex Hologram)
This method uses multilayer colored film as a recording medium to obtain most of the advantages

of the Kinoform without its major disadvantages. Both the phase and amplitude are recorded. Using Kodachrome film, the intensity-variation pattern is exposed through a red filter and the phase-variation pattern is exposed through a cyan filter. If the processed film is illuminated with a HeNe laser, the cyan layer modulates the amplitude and the other two layers mod- ulate the phase.

Advantage: Since all the light is diffracted into a sin- gle image, the diffraction efficiency of the ROACH is very high. In addition, because both the amplitude and phase information is recorded, the image quality is superior to that of the Kinoform. The ROACH is supe- rior to the Detour Phase Hologram because only one display spot is required for each Fourier coefficient and quantization noise is negligible.

Disadvantage: The steps required to produce this hologram are much more involved than for the Kinoform .

Computer-Generated Interferograms
Problems were encountered with the Detour Phase Holograms when encoding wavefronts with large phase variations. These problems were attributable to the fact that the apertures in the cells overlapped in cases where the phase of the wavefront moved through a mul- tiple of 21st radians or more. To solve this, an alternative approach was taken by noting that the case of a wavefront which has no amplitude variations is essentially similar to an interferogram. The non-linearity of the emulsion can then be exploited to produce a hologram that is approximately binary. There are methods that can then be used to record the amplitude variations in the binary fringe pattern.

Advantage: Computer-generated interferograms are an improvement to the Detour Phase Hologram where large phase variations are encountered.

Disadvantage: Since amplitude variation is not recorded initially, it has to be calculated or derived later by one of several means.

Three Dimensional Computer-Generated Holograms
Computer-generated holograms were first generalized to a three-dimensional object in 1968 by Waters. The process involved approximating the 3D object by making a number of equally spaced cross sections perpendicular to the z axis. Then a number of holograms from different angles were produced showing the resulting changes in parallax. Due to the fact that there were lines "hidden" from the front view of the object encountered as successive slices through the object were made,

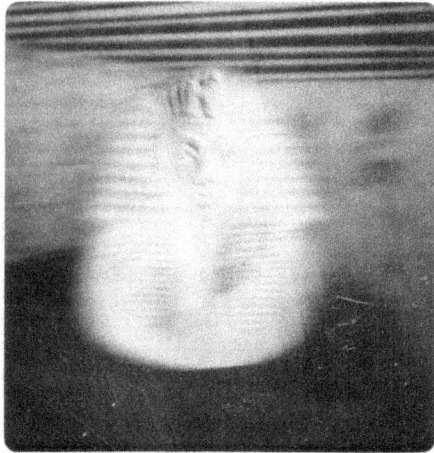

one had to add the contributions to the object wave arising from the hidden lines as one went along.

In 1970, King, Knoll and Berry took a different approach. Their technique makes a tall, thin, holo- graphic stereogram and then "multiplexes" or joins large quantities of these stereograms together on a sin- gle plate. The computer produces a series of perspective projections of an object by either programming the holograms or by filming the subject from a number of different perspectives along a horizontal axis, and inputting the holograms to the computer by microdensitometer readings. This is a nice theory, but reports are that it is difficult in practice.

The computer then outputs a series of thin vertical strips, each of which are holograms, on a single plate. Since this is a thin hologram, it can be illuminated with white light to construct a bright, almost achromatic image. When the hologram is illuminated by the reference beam, or any white light, we see the real image, which is two-dimensional and located in the plane of the final hologram. However, since our eyes see a number of different holograms and each one is from a slightly different perspective, the viewer has a stereoscopic view and sees a three-dimensional image. If the plate is large and there are numerous images, we can tilt the plate from side to side and see the image move about.

Computer-Generated Holograms in the Marketplace: A Case History

One recent successful implementation of the technique just discussed is "A Clear Day", a CGH produced by Starlight Holographics of Ottawa, Canada and Applied Holographics of Oxnard, California, USA. The project was commissioned by Northern Telecom to celebrate its status as the first international telecommunications company to stop using the ozone-depleting compound CFC-113; hence the name "A Clear Day".

The hologram was created by modeling photo- graphic images of Thomas Van Sants' original work, The Earth from Space . It was done with Alias Software on an Iris Computer system. The imported database took approximately five days of non-stop computer ren- dering time to create the three hundred perspectives that were necessary to create the final three-dimensional image of the earth.

After computer input, three hundred frames of film were shot from the image on a high-resolution monitor. Then, in Applied Holographies' optics lab, laser light was passed through each individual frame onto the recording plate, resulting in a master hologram, of

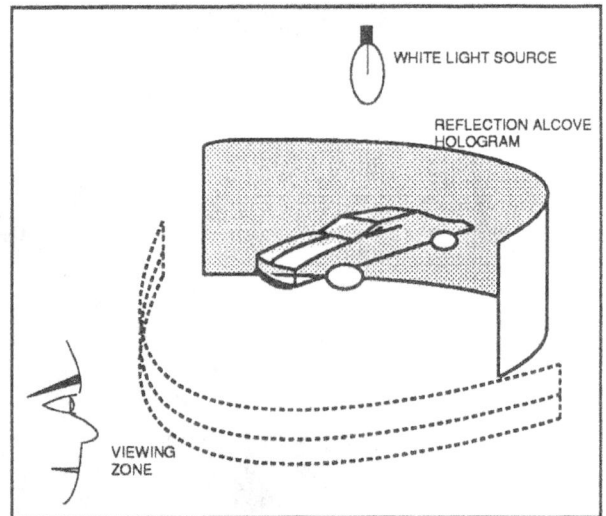

Alcove hologram

which 130 copies were made (70 for Northern Tele- com offices worldwide).

The final large format hologram shows three per- spectives of the earth (100 frames for each perspective), with every country in the world included. The final size of the 130-print run was 32" x 45", although Starlight president Stephen Leafloor notes that a smaller commercial embossed run size (6" x 6") will begin this year. Starlight says that this is the first hologram created solely from satellite data.

"A Clear Day" was on display at the Earth Sum- mit in Rio De Janiero in 1992, and will also be on display at the Smithsonian National Museum in Washington, DC, USA and the National Museum of Science and Technology in Canada.

Alcove Hologram

The alcove hologram is the latest (Benton, 1987) in the evolution of 3D computer-generated holograms. This is a multiplexed reflection stereogram in which the multiplexed holograms are arranged in a semi-circular alcove into which the viewer looks.

The advantage is that as the viewer walks from side to side, the three-dimensional hologram rotates left to right, allowing the viewer to see around the object. To date, a viewing angle of about 30 degrees has been accomplished. Benton suggests enlarging and extending the arrangement so as to have close to 180 degrees of viewing using 900 slit holograms that are 300 mm high and 1mm wide (the interior of the concave would be 600 mm across).

Advantage: Although a volume hologram is the only true three-dimensional recreation of an object, it has a problem with the restricted angle of view and subject matter. Holographic stereograms use any movie film that is filmed with the proper perspectives and, recently, color has been used in holographic stereograms. Industry benefits include being able to move around any desired object at will with out having the object present.

Disadvantage: One of the big problems with the CGH is getting the holographic image into the computer. Using a microdensitometer to read 900 holograms into the computer is more of a challenge than anyone wants to take on. Generating the hologram internally by programming the wave patterns is very difficult for all but the simplest objects. Progress is being made, however, and developing technology is on the side of the multiplexed CGH. It should be pointed out that curved, multiplexed holograms made without the aid of the computer are commercially available, but the computer will offer enormous benefits when the system is perfected.

Uses of Computer-Generated Holograms

The uses for CGHs and HOEs are very similar, since one of the main values of the CGH is its ability to produce almost noise-free HOEs.

Some applications for CGHs:

- *Commercial Embossing:* Used to make simple artistic holograms for later use embossing, etc.
- *Multiplexed Displays:* 3D artistic displays using holographic stereograms such as the alcove hologram.
- *Medicine:* Currently under research at MIT is a process that takes CAT scans and MRI (magnetic resonance imaging), both already digital, and cre- ates a multiplexed CGH from the data.

Footnote:
1. *Thick And Thin Holograms:* There are two popular mathematical models used to describe the pattern of light waves coming through a hologram. One model, the Raman-Nath diffraction pattern, is based on the assumption that the thickness of the emulsion is small compared to the average spacing of the interference fringes. The other model, called the Bragg diffraction pattern, is used to describe results when the thickness of the emulsion is large compared to the average spacing of the interference fringes. There is a "boundary" between the two models where neither model holds up perfectly. Therefore, one has to say that the Raman-Nath "thin hologram" model works when the emulsion thickness is considerably less than the fringe spacing and the "volume hologram" Bragg model works when the thickness of emulsion is considerably greater than the fringe spacing.

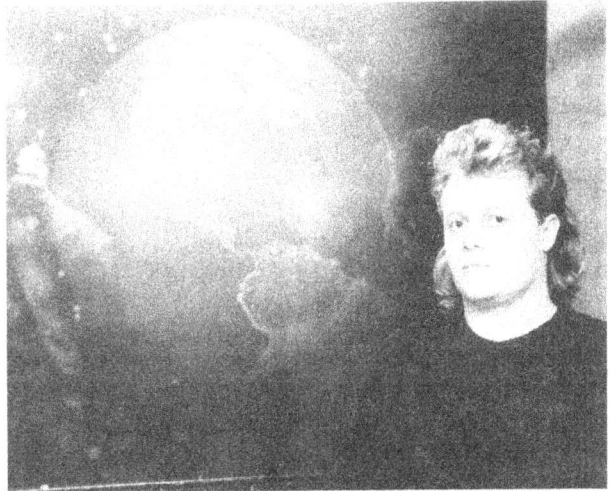

Stephen Leafloor with "A Clear Day"

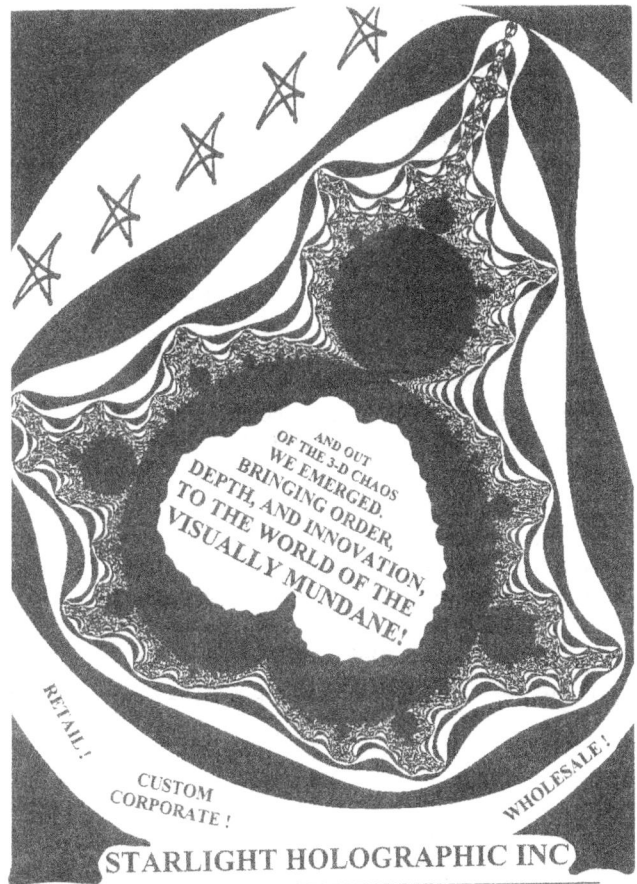

AND OUT OF THE 3-D CHAOS WE EMERGED. BRINGING ORDER, DEPTH, AND INNOVATION, TO THE WORLD OF THE VISUALLY MUNDANE!

RETAIL!

CUSTOM CORPORATE !

WHOLESALE!

STARLIGHT HOLOGRAPHIC INC

References & Reading:

For Binary Detour Phase Holograms:

Brown, R.B. & Lohmann, A.W. (1966). Complex Spatial filtering with binary masks. Applied Optics, 5, %7-9.

_____. (1969). computer-generated Binary Holo- grams. IBM Journal of Research & Development, 13, 160-7.

Burckhardt, CB. (1970). A simplification of Lee's method of generating holograms by computer. Applied Optics, 9, 1949.

Dallas, W.J. (1971a.) Phase quantization - a compact derivation. Applied Optics, 10, 674-6.

_____. (1971b.) Phase quantization in holograms - a few illustrations, Applied Optics, 10, 674-6.

Goodman, J.W. & Silvestri, A.M. (1970). Some effects of Fourier domain phase quantization. IBM Journal of Research & Development, 14,478-84.

Haskell, R.E. & Culver, B.C. (1972). New coding technique for computer-generated holograms. Applied Optics, 11,2712-14.

_____. (1973). Computer-generated binary holograms with minimum quantization errors. Journal of the Optical Society of America., 63, 504.

Lee, W.H. (1970). Sampled Fourier transform hologram generated by computer. Applied Optics, 9,639-43.

Lohmann, A.W. & Paris, D.P. (1967). Binary Fraunhofer holograms generated by computer. Applied Optics, 6, 1739-48.

References for The Kinoform:

Kermisch, D. (1970). Image reonstruction from phase information only. Journal of the Optical Society of America, 60, 15-7.

Lesem, L.B., Hirch, P.B., & Jordan, J.A. (1%9). The Kinoform: A new wavefront reconstruction device. IBM Journal of Research & Development, 13,150-5.

Lohmann, et alia. Binary Fraunhofer holograms gener- ated by computer. op.cit., pp.1739-48.

References for the R.O.A.C.H.:

Chu, D.C., Fienup, J.R. & Goodman, J.W. (1973). Multi-emulsion, on-axis, computer-generated holograms. Applied Optics, 12, 1386-8.

References for Computer-generated Interferograms:

Bryngdahl O. and Lohmann, A.W. (1968) Inter- ferograms are image holograms. Journal of the Optical Society of America, 58,141-2.

Lee, W.H. (1974). Binary synthetic holograms. Applied Optics, 13, 1677-82.

_____. (1979). Binary computer-generated holo- grams. Applied Optics, 18,3661-9.

References for Three Dimensional CGH:

Benton, S.A. (1982), Survey of holographic stereograms, SPIE Proceedings., 367,15-19.

_____. (1987), Alcove Holograms for Computer- Aided Design, SPIE Proceedings, 761, 53-6l.

Brown, et alia. Computer-generated binary holograms. op.cit. pp. 160-7.

Holzbach, M., (Sept. 1986), Three Dimensional Image Precessing for Synthetic Holographic Ste- reograms, M. Sc. thesis, Massachusetts Institute of Technology.

King, M.C., Noll, A.M. & Berry, D.H. (1970). A new approach to computer generated holography. Applied Optics, 7,1641-2.

Krantz, E. (Sept. 1987), Optics for Reflection Holographic Stereogram Systems, M. Sc. thesis, Massachusetts Institute of Technology.

Lesem, et alia. The kinoform: a new wavefront reconstruction device. op. cit. pp.150-5.

Teitel, M. (Sept. 1986), Anamorphic Ray Tracing for Synthetic Alcove Holographic Stereograms, M. Sc. thesis, Massachusetts Institute of Technology,

Waters, J.P. (1968). Three-Dimensional Fourier transform method for synthesizing binary holo- grams. Journal of the Optical Society of America, 58, 1284-8.

9

Holographic Non-Destructive Testing

Want to test something's ability to withstand pressure without breaking it into bits? Want to study incredibly small movements in a liquid? These practices and others come under the discipline of Holographic Non-Destructive Testing (NDT).

NDT is a rapidly-growing field of holography. Although the techniques and analysis can be very involved, the general concept is not difficult to understand.

Suppose you make a transmission hologram of a still object. Then, without disturbing the setup, you make another transmission hologram of the object. If you were to take the two holograms and put them on top of each other, you would simply see one hologram pattern reinforced by the other, since both holograms are identical. If there is a small amount of movement in the object between exposures, however, the two patterns would be slightly different at the site of the dislocation.

When you put the two patterns on top of each other, you then see a Moire pattern where they differ. These are called interference patterns. Since the holograms do not match at the site of the dislocation, the interference patterns show us where, and by how much, the object is dislocated. Investigations that started in 1965 showed that in some cases, very exact measurements can be obtained (measurements down to the order of light waves in some cases) by studying the interference patterns.

The immediate applications are clear. You can take one holographic exposure of an object, subject the object to stress, and then take a second holographic exposure. Comparing and analyzing the interference patterns that result when the two patterns are superimposed tells you where the object deformed under stress. Ana- lyzing this has become a whole industry in itself and is variously called "Holometry", "Holographic Interferometry", or by its much more sales-oriented name of "Holo- graphic Non Destructive Testing" (NDT).

Applications

- Locating the presence of a structural weakness: e.g. the object is stressed by the application of a load or change in pressure or temperature.

- Detecting cracks and the location of areas of poor bonding in composite structures.

- Medical and dental research.

- Aerodynamics, heat transfer, and plasma diagnostics.

- Solid mechanics, such as measuring the changes in shape due to absorption of water and corrosion.

There are a number of techniques for holographic interferometry testing, each with its advantages and disadvantages. Some of the major classifications follow.

Real-time Holographic Interferometry

This test is done to obtain immediate results. A hologram is shot, developed and then replaced in the holder it was in for the exposure. When the object is illuminated as if to expose it for another shot, there are two holographic patterns on the plate. One pattern will be the one you developed, while the second is created by the light you just turned on, reflecting from the object still in its original position.

The two patterns should match exactly-- if nothing moved and the plate is replaced exactly in its original spot. Therefore, as you look through the plate at the illuminated object, you should see just one reinforced holographic pattern.

Suppose, however, that there was some small movement in your object while you were developing the plate. If there was movement, then after the plate is put back in the holder and the laser is turned on, the observer viewing the object sees it covered with a pattern of interference fringes around the areas where the object has changed shape. If you move the object or put it under stress while viewing it, you see interference fringes change to map the area where the object is being deformed.

It is awkward to have one person at a time viewing the interference fringes. Capturing the image being viewed is also a problem. These problems may be solved by using a closed circuit television camera. A Polaroid camera can photograph any desired scene. It is then possible to color-code the fringes to identify the direction of the displacement by using filters and a double exposure. Another method is to have the video signals read digitally and input directly to a computer for analysis.

Since pulsed lasers are not subject to vibration restrictions, it is also possible to mount a pulsed laser on a truck, drive to an object that cannot be moved, such as a water tower, and from the truck perform a real-time NDT stress analysis. This also applies to buildings, bridges, etc. for applications like earthquake analysis.

Although the above procedure for real-time holographic interferometry sounds very simple and straightforward, there are many problems with it. Precise positioning of the hologram after processing is necessary, and uniform drying of the emulsion to avoid deformations in the hologram is another potential problem. One method of solving these problems is to develop the hologram on the spot using a "liquid gate" arrangement.

Another alternative for processing the hologram, which eliminates the need for wet processing altogether, is to use thermoplastic recording. There are commercial units available in which a hologram can be recorded and viewed in less than a minute.

Advantage: The advantage of real-time NDT is its instant results. There are numerous cases where you must have instant analysis so that work on a project can continue. You are also able to view the interference patterns as the object is undergoing stress. In other words, you see a real-time "movie" of the stress points developing as the experiment is conducted, and you may photograph or videotape what you see.

Disadvantage: One disadvantage is that there is a serious drop in the visibility of fringes because the light diffracted by the hologram remains linearly polarized, while the light scattered by the object is largely depolarized. To avoid the decline in fringe visibility, it is necessary to use a polarizer when viewing or photographing the fringes. A second disadvantage is exact registration of a developed hologram with the original object. A third disadvantage is that you do not wind up with a precise, permanent record of the interference fringes except for what appeared on the television monitor. Recording on photographic emulsion offers more precision.

Double Exposure Holographic Interferometry

In this case a double exposure of the hologram is made on the same plate. The first exposure is usually made with the object in an unstressed condition, and the second with stress applied to the object.

Advantage: Double exposure interferometry is much easier to perform than real-time holography because the two holograms are in exact register. Any distortions to the emulsion affect both holograms equally. Both holograms have the same polarization. Therefore, the fringes are much clearer and no special care needs to be taken when illuminating the hologram. It is also possible to use rainbow holograms for double exposure interferograms. Fringe patterns due to different effects can then be displayed in different colors.

Disadvantage: The double exposure tends to brighten the image of the object, which makes it difficult to see small displacements. You can help correct this by shifting the phase of the reference beam or tilting the object beam between exposures. Another disadvantage is that with a double exposure you have the "before" and "after" snapshot but you do not see what happens in between. This can be overcome, to some extent, by multiplexing techiques. Another problem is that you cannot compensate for and control the fringes if the object tilts between exposures. This problem can be solved by using two holograms, recorded on the same plate, with two different, angularly-separated reference waves.

Sandwich Holographic Interferometry

The sandwich hologram takes advantage of the fact that multiple exposures can be made on the same holographic plate. This procedure provides a much more elegant solution to the problem of viewing varying stages of stress on an object between ex posures.

Two plates (with no anti-halation backing) are set in the same plate holder with their emulsion surface toward the object. An exposure is made of the object under no stress. The object is then tilted, stress is applied, and a second exposure is made. This procedure is repeated as many times as desired.

To view the hologram, you put the plate that recorded the undisturbed object in its original plateholder, and add to the plateholder one of the holograms of the object under stress. You see interference patterns at the points of deformation. You can see the incremental deformations of your object by going through all the plates in sequence. Using any combination of two plates, the incremental changes between any two stages of deformation can be seen.

Advantage: This technique allows you to see a wide latitude or combination of stress loads. You- could, for example, select the plate B1 and F2 to view what happens when the first stress load is applied. Or, you could select B2 and F3, which show the deformations between stress load 1 and 2.

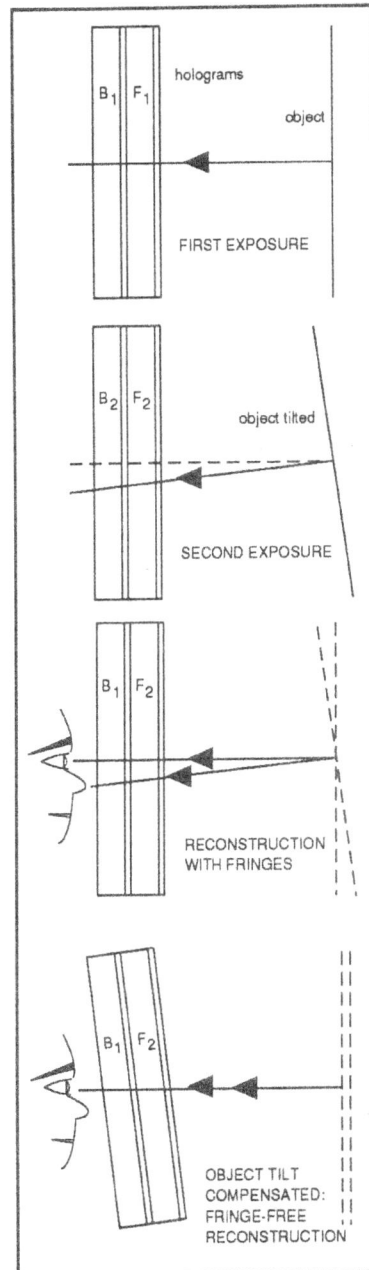

Sandwich hologram interferometry

Disadvantage: You are looking through two different plates to see the image and there may be some parallax problems. It is also time- and material-consuming to catalogue and handle all these different plates.

Holographic Interferometry Through Dense Media

One example of this procedure begins when a hologram is taken of an undisturbed chamber filled with a gas. The chamber is disturbed in some way and a sec- ond hologram is taken. Any variation in density of the gas alters the path length of the laser light. This results in a hologram which produces interference pat- terns when compared to the undisturbed chamber's hologram.

Another example of dense-media interferometry is plasma diagnostics, in which measurements of the light's refraction at two wavelengths make it possible to determine the electron density directly.

Time-A veraged Holographic Interferometry

There are numerous instances in which the object can only be properly tested while operating. To see clearly all the uneven stress in a hi-fi speaker or a rotating fan, for example, you need to test them while they are in operation. In order to make interferometer tests on these subjects a number of clever methods have been developed. Time-averaged holographic interferometry is one of them.

If you take a flexible metal ruler and whip it back and forth in the air while holding the end of it tightly in your fist, you see the image of the ruler at the two most extreme points of the flex, but it is a blur in between. If you were to calculate it, you would find that the amount of time spent at the two extremes of the cycle is a substantial percentage of the time in the entire cycle.

Suppose we look directly into a hi-fi speaker. When it is operating, the speaker membrane or cone spends most of its time at the two ends of its vibration (i.e. the cone is either fully extended toward us or fully extended away from us). We simply make a hologram of the speaker while it is operating and extend the exposure so that it is long compared to the cycle of the speaker membrane moving in and out.

What is the result? Since the time spent between the two extremes is a blur, the result is a hologram of the speaker membrane fully extended toward us, superimposed on a hologram of the speaker membrane fully extended away from us. In other words, we seem to obtain a "double exposure hologram" even though it is only one exposure.

If the speaker is not in perfect resonance and the distance the speaker membrane extends towards us is not exactly equal to the distance the speaker membrane moves away from us, the two superimposed holographic patterns create an interference pattern. You can calculate how out of resonance the speaker is by counting the fringes. The distance between fringes is slightly greater than one quarter of a wavelength.

The above technique of time-averaged holographic interferometry works well for studying objects vibrating in a stable manner. It was first reported in 1965 and is usually used to identify the resonances of a test object. It does this by monitoring the object's response while varying the excitation frequency and the point of excitation on the object.

Typical applications of the process are testing of musical instruments and electromechanical devices such as loudspeakers, turbine blades and aircraft structures.

Advantage: In time-averaged holographic interferometry you are allowed to view and adjust your object to obtain an almost perfect resonance with little effort.

Disadvantage: Time-averaged holographic interferometry is good only for objects vibrating in a stable manner, and compares only the two extremes of the vibration cycle. This procedure does not show the relative phases of the various modes of vibration. Also, the contrast of the fringes falls off rapidly as the fringe order increases.

Strobed Holographic Interferometry

Suppose that you have an object that vibrates in a stable manner, but the point you want to inspect in the vibrating cycle is not at one of the two extremes of the cycle. In 1968, articles were written describing a solution to this problem using a sort of strobe light effect. In this process, a sequence of light pulses are triggered at desired times during the cycle. The resulting hologram is equivalent to a double exposure hologram (or sandwich hologram if you use that method) recorded while the object is in the desired state of deformation at any time in the cycle.

Advantage: Allows you to study the object at any time in the vibration, in relation to the object at any other time in the vibration.

Disadvantage: This is a much more involved set-up than the time-averaged technique.

Holographic PIV Vector Map

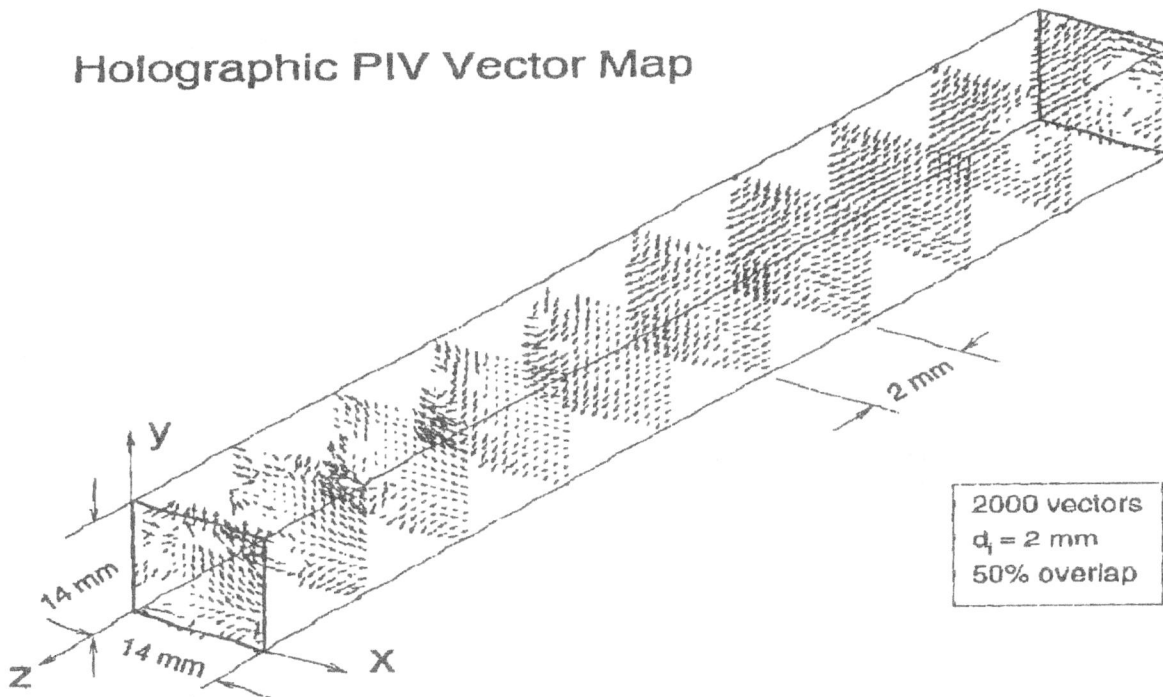

2000 vectors
d_i = 2 mm
50% overlap

Interferometry in the Lab: Holographic Particle Image Velocimetry

One example of NDT through dense media (discussed on the previous page) is Holographic Particle Image Velocimetry (HPIV). Donald Barnhart at the University of Illinois/Urbana's Department of Electrical and Computer Engineering, is one person doing research using HPIV, and offered the following explanation.

"Holographic techniques for measuring particles have been studied since the inception of holography. However, techniques used in the past have had limited success in particle velocimetry for one of the following reasons: large depth of focus, poor signal-to-noise ratio, distortion, and directional ambiguity in the particle displacement.

In the context of fluid mechanics, volcimetry is the measurement of vector flow fields. Understanding how fluids interact with themselves and surrounding solid structures is of great concern in the design of automobiles, chemical processing facilities, airplanes, rockets, turbomachinery, and in understanding the physics of complex flows.

Computer solutions of fluid flows are difficult even on today's fastest supercomputers, but they provide a large amount of information about the flow In the form of large three dimensional vector fields. The thrust of modern experimental flow velocimetry is to provide comparable information by measuring instantaneous velocity vectors throughout an entire fluid flow.

Particle velocimetry involves measuring the movements of tiny particles suspended in the fluid flow environment. If the particles are sufficiently small and neutrally buoyant, they will be carried by the fluid and will not significantly change the fluid's characteristics. These particles scatter light in the otherwise transparent fluid environment, acting as markers for the instantaneous positions within the fluid volume. By obtaining the particle positions at two points in time, the particle displacement in space is known. Given the particle displacement and the time delay for the displacement, the velocity of the particle is known. Hence the velocity of the fluid volume surrounding the particle is obtained.

In holographic particle image velocimetry (HPIV), two or more holograms are made of the fluid volume at two or more distinct times. The holographic images contain information about the spatial positions of the particles within the volume being measured. By measuring the three-dimensional change in position of the particle images, the three dimensional velocity at each point in the fluid volume can be determined.

Particle imaging was one of the first applications for holography. Early researchers such as B.J. Thompson wrote numerous papers on using in-line holographic technique for particle analysis. Over the last decade, methods for taking two-dimensional particle image velocity measurements using conventional photography have become a relatively standard procedure. A num- ber of efforts to perform holographic particle image velocimetry have been reported.

We (at University of Illinois/Urbana) are presently developing a three-dimensional velocimetry system that combines existing technology for analyzing 2-D images with holographic recording. The goal is to obtain a three-dimensional velocity field map for a vol- ume of (100 mm) 3 at regular intervals of 1mm3, giving a total of 1,000,000 three-dimensional velocity vectors using 107 particles (- 10 particlesivoxel)."

As one might expect, Barnhart's team uses a very powerful pulse laser (frequency doubled Nd:YAG). The system combines four lenses and two prisms behind the volume being measured. The laser sends out pulses of 300mJ. In one recent experiment, a vector flow was reconstructed using 28 different "slices" of .5mm length (120,000 vectors). From this information a three-dimensional Holographic PIV Vector map can be constructed.

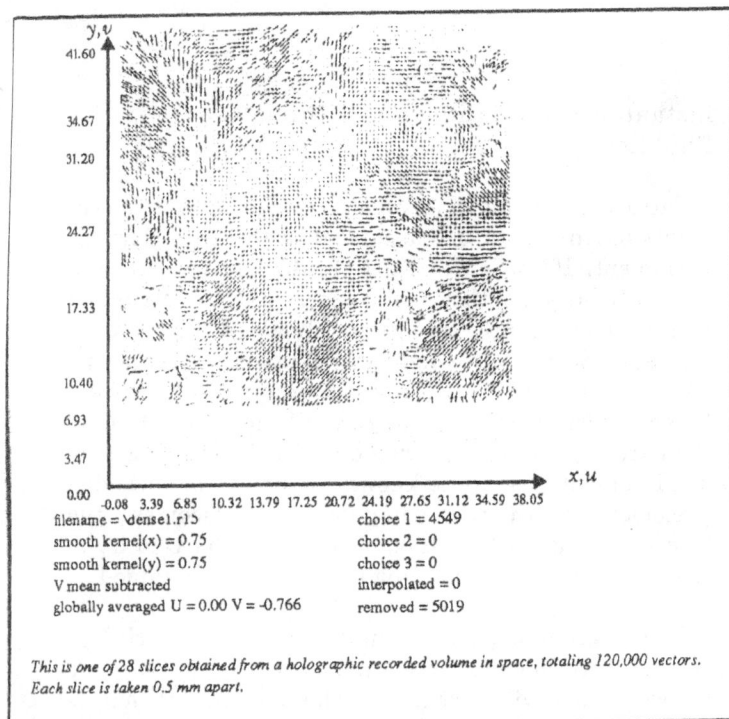

filename = \dense1.rl5   choice 1 = 4549
smooth kernel(x) = 0.75   choice 2 = 0
smooth kernel(y) = 0.75   choice 3 = 0
V mean subtracted   interpolated = 0
globally averaged U = 0.00 V = -0.766   removed = 5019

This is one of 28 slices obtained from a holographic recorded volume in space, totaling 120,000 vectors. Each slice is taken 0.5 mm apart.

References & reading:

General:
Hariharan, P . (1984) Optical Holography. Cambridge: Cambridge University Press.

References for applications of holographic interferometry:
Greguss, P. (1975). Holography in Medicine. London: IPC Press]

_____. (1976). Holographic interferometry in bio- medical sciences. Optics & Laser Technology, 8, 153-9.

Sciammarella, C.A., (1987). Advances in the appli- cation of holography for NDE, SPIE Proceedings, 523, 137-44.

Von Bally, G. ed. (1979). Holography in Medicine and Biology. Berlin: Springer-Verlag.

Willenbourg, G.C., (1987). Holography and Holometry Applications in Dental Research, SPIE Proceedings, 747, 51-6.

References for Interference Fringes:
Powel, L. & Stetson, K.A. (1965). Interferometric Vibration Analysis by Wavefront reconstruction, Journal of the Optical Society of America, 55, 1543-8.

References for real-time holographic interferometry:
Dandliker, R., Ineichen, B. & MOltier, F.M. (1973). High resolution hologram interferometry by electronic phase measurement. Optics Commu- nications, 9, 412-16.

Hariharan, Oreb & Brown, (1982). A digital phase- measurement system for real-time holographic interferometry. Optics Communications, 41 , 393- 6.

_____. (1982). Real-time holographic interferometry: a microcomputer system for the measurement of vector displacements. Applied Optics, 22, 876- 80.

References for liquid gate recording:
Hariharan, P. (1977). Hologram interferometry: identification of the sign of surface displacements. Optica Acta, 24, 989-90.

Hariharan, P. & Ramprasad, B.S. (1973). Rapid in situ process. for real-time holo. interferometry. Journal of Physics E: Sci. Instruments, 6, 699-701.

Van Deelen, W. & Nisenson, P. (1969). Mirror blank testing by real-time holographic interferome- try. Applied Optics, 8, 951-5.

References for thermoplastic recording:
Hariharan, P. & Hegedus, Z.S. (1975). Relative phase shift of images reconstructed by phase and amplitude holograms. Applied Optics, 14,273-4.

Saito, T., Imamura, T. & Tsujiuchi, J. (1980). Solvent vapor method in thermoplastic photo conductor media. Journal of Optics (Paris), 11,285-92.

Thinh, V.N. & Tanaka, S. (1973). Real-time interferometry using thermoplastic hologram. Japanese Journal of Applied Physics, 12, 1954-5.

**References for double exposure
holographic interferometry:**
Yu, F.T.S. & Chen, H. (1978). Rainbow Holo- graphic interferometry. Optics communications, 25, 173-5.

Yu, F.T.S., Tai, A. & Chen, H. (1979). Multiwavelength rainbow holographic interferometry. Applied Optics, 18,212-18.

_____. (1979). Multislit one-step rainbow holographic interferometry. Applied Optics, 18,6-7.

References for shifting reference beams:
Collins, L.F. (1968). Difference holography. Applied Optics, 7,203-5.

Hariharan, P. & Ramprasad, B.S. (1972). Simpli- fied optical system for holographic subtraction. Journal of Physics E: Scientific Instruments, 5, 976- 8.

_____. (1973). Wavefront tilter for double-expo- sure holographic interferometry. Journal of Physics E: Scientific Instruments, 6, 173-5.

Jahoda, F.e., Jeffries, R.A & Sswyer, G.A. (1967). Fractional fringe holographic plasma interferometry. Applied Optics, 6, 107-10.

References for multiplexing techniques:
Caulfield, H.J. (1972). Multiplexing double-expo- sure holographic interferograms. Applied Optics, 11,2711-12.

Hariharan, P. & Hegedus, Z.S. (1973). Simple multiplexing technique for double-exposure hologram interferometry. Optics Communications, 9, 152-5.

Parker, R.J. (1978). A new method offrozen-fringe holographic interferometry using thermoplastic recording media. Optica Acta, 25, 787-92.

Two holograms on the same plate:
Ballard, G.S. (1968). Double exposure interferometry with separate reference beams. Journal of Applied Physics, 39,4846-8.

Gates, J.W.C. (1968). Holographic phase recording by interference between reconstructed wavefronts from separate holograms. Nature, 220, 473-4.

Tsuruta, T., Shiotake, N. & Itol, Y. (1968). Hologram interferometry using two reference beams. Japanese Journal of Applied Physics, 7, 1092-100.

**References for sandwich
hologram interferometry:**
Abramson, N. (1974). Sandwich hologram interferometry: a new dimension in holographic comparison. Applied Optics, 13,2019-25.

_____. (1975). Sandwich hologram interferometry: 2. Some practical calculations. Applied Optics, 14,981-4.

_____. (1976). Sandwich hologram interferometry: 3: Contouring Applied Optics, 15,200-5.

_____. (1976). Holographic contouring by translation. Applied Optics, 15, 1018-22.

_____. (1977). Sandwich hologram interferometry: 4: Holographic studies of two milling machines. Applied Optics, 16,2521-31.

Abramson, N. & Bjelkhagen, H. (1978). Pulsed sandwich holography. 2. Practical application. Applied Optics, 17, 187-91.

Hariharan, P. & Hedgedus, Z.S. (1976). Two-hologram interferometry: a simplified sandwich technique. Applied Optics, 15,848-9.

**References for holographic
interferometry through dense media:**
Ostrovskaya, G.V. & Ostrovskii, Yu. I. (1971). Two-wavelength hologram method for studying the disper- sion properties of phase objects. Soviet Physics - Tech- nical Physics, 15, 1890-2.

Radley Jr, RJ. (1975). Two-wavelength holography for measuring plasma electron density. Physics of Fluids, 18, 175-9.

Zaidel, AN., Ostrovskaya, G.V. & Ostrovskil, Yu. I.(1969). Plasma diagnostics by holography. Soviet Physics - Technical Physics, 13, 1153-64.

**References for time averaged
holographic interferometry:**
Agren, e.H. & Stetson, K.A (1972). Measuring the resonances of treble viol plates by hologram interferometry and designing an improved instrument. Journal of the Acoustical Society of America, 51, 1971-83.

Bjelkhagen, H. (1974). Holographic time average vibration study of a structure dynamic model of an airplane fin. Optics & Laser Technology, 6, 117-23.

Chomat, M. & Miller, M. (1973). Application of holography to the analysis of mechanical vibration in electronic components. TESLA Electronics, 3, 83-93.

Saxby, G., (1988). Practical Holography, Prentice Hall, 321.

Stetson KA. & Powell, RL. (1965). Interferometric hologram evaluation and real-time vibration analysis of diffuse objects. Journal of the Optical Society of America, 55, 1694-5.

References for strobed holographic interferometry:
Archbold, E. & Ennos, A. E. (1968). Observation of surface vibration modes by stroboscopic hologram interferometry. Nature, 217,942-3.

Shajenko, P. & Johnson, C.D. (1968). Stroboscopic holographic interferometry. Applied Physics Letters, 13,44-6.

Watrasiewicz, B.M. & Spicer, P. (1968). Vibration analysis by stroboscopic analysis. Nature 217, 1142-3.

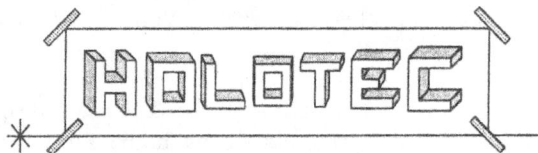

10
Embossed Holograms

Embossed holograms are the easiest to mass-produce, and have proven effective in a number of promotional and security applications. They are by far the most commonly used holograms. This chapter explains how they are made and applied to products; what the costs are; and the range of applications for embossed holography.

The Cheapest Way To Go

Embossed holograms are holograms which are mass-produced by taking a shim, or metal "negative" of the holographic image (interference pattern), and making impressions of the image onto a desired substrate: foil is probably the most popular due to its low cost, but materials as diverse as chocolate have been used for embossed holograms! There is no question that embossed holograms are by far the cheapest way to mass-produce holograms.

The major drawback to embossed holograms is that they do not have a lot of depth. Embossed 3D holograms tend to "smear" after a certain depth (1").

There are great advantages to embossed holograms, however, and there are tricks one can use to get around the problem of depth. For example, since a photograph is 2D and has no depth it is an ideal subject. Furthermore, there is no reason why you can not take several photographs, splice them together in extremely small strips and produce a three-dimensional effect for the viewer. This can even be done in color. An example of this process is the Michael Jordan hologram that appeared on the cover of *Sports Illustrated,* one of America's most popular magazines. Here is the story of how that was done:

Michael Jordan hologram

Upon selecting Michael Jordan as their Sportsman of the Year for 1991, Sports Illustrated decided to use a holographic stereogram of the basketball star on its year-end cover.

Sharon McCormack made the original master hologram by seating Jordan on a turntable in a Chicago studio. The turntable was rotated 120 degrees at approximately 1 1/3 r.p.m., and Jordan was cued to smile (very slowly!) as he spun. A 200-frame strip of film (ten seconds' worth) was chosen by McCormack and sent to American Bank Note Holographics for mass pro- duction. The total production run was 4.1 million holograms (foil labels) which were applied to *Sports Illustrated's* cover.

Applying Embossed Holograms

The press embosses the holograms on rolls of stock which, in the most common commercial work, have an image area of six square inches for each impression. The roll consists of a continuous ribbon of 6 x 6 inch embossed squares separated from one another by about one-half inch. The holograms are often arranged on each 6 x 6 inch square as rine, 2-inch squares ("Nine- up") of different holograms, but you can use the entire 6 x 6 inch square for one hologram.

Stickers vs. Hot stamping

There are a number of options for displaying your final embossed product. Embossed holograms can be produced as stickers cut to your size specifications. Typical of the final holograms are individual peel-and- stick holograms. For stickers, be sure the substrate chosen is thick and strong enough not to conform to a textured surface onto which it might be applied. The best surface on which to stick a hologram is smooth and rigid, so you can be sure that the hologram is flat and consequently able to reconstruct its image properly. There are applicator machines to apply the sticker to your product, but it is often done by hand.

Instead of stickers, you may hot-stamp the foil directly onto a surface like a book cover. In general, the hot foil application keeps precise registration bet- ter than methods of sticker application. Although hot-stamping appears more expensive, when you consider the costs of a very large run using sticker application they both can come out about equal in cost. Smaller runs seem to favor sticker application.

If you choose hot-stamping, be absolutely certain of the strength and smoothness of the cover material. The embossed foil is very thin---like the foil around individual sticks of chewing gum. If the hologram has ripples in it as a result of a bumpy surface, the image itself might appear rippled.

It is a good idea to have the surface approved by the holographer and the printer, together, before going into production. Also beware of coated paper stock and printed surfaces. These surfaces may present prob- lems for adhesion of holograms. Regardless of which method you choose, it is advisable to be there when the holograms are applied so you can check the quality.

Hot Stamping & costs:

As we just mentioned, embossed holograms can come either as self-adhesive stickers, or as foil that is hot-stamped onto your product. The preferred method for very long runs is hot stamping, although the price of stickers has become competitive in recent times. Hot-stamped holograms can be applied with more pre- cision, and it is easier to integrate them into the design of a piece of graphic art.

Two Hot-Stampers Talk Shop

Says Mark Mulvaney of Letterhead Press in Wisconsin, USA, a major hot-stamping company, "the differences between stickers and hot-stamping are in speed, quality and turn-around time. Clients use hot- stamping when the image is to be a more integral part of the overall design. Placement is better (with hot-stamping than with stickers). The cost is about the same. Speed is a little slower (with hot-stamping than with stickers). Capacity for the presses is growing."

Mulvaney points out that Letterhead frequently uses 40" press sheets, and hot-stamped the cover of Prince's latest CD "9-up" on a 40" press. Letterhead also recently had a 30" press installed, which indi- cates that there is growing capability to handle vari- ous size foil sheets, too.

Hank Brandtjen of Brandtjen & Kluge, a manufacturer of hot-stamping machinery, had this to say about the hot-stamping field.

"The challenges of holography are fading and its use is becoming more practical and affordable. The economic principles associated with supply and demand have been at work, and today producing a piece with a hologram on it is a more affordable alternative.

While other processes do exist, the most common, accessible and low-cost holographic medium for incorporation into print is the hot-stamping of holographic foil to the final stock. Towards this end, producing a piece with a hologram on it is the culmination ofthree necessary steps. You need a foil stamping and embossing press, a hologram registration unit, and hologram foil.

Foil stamping and embossing presses fall into four main categories: platen automatic sheet fed, cylinder automatic sheet fed, continuous pack-to-pack, and hand-fed. While nearly every press can be used to stamp hologram foil, some are clearly more widely used. Products such as cereal boxes, magazine covers and cosmetic packages are produced on platen auto- matic sheet fed presses. Security documents and checks produced on a continuous web are produced on continuous pack-to-pack presses.

Holograms can be differentiated by being either random pattern or set image. An example of a random pattern hologram is the continuous holographic pat- tern found on a BacardiThl rum box. An example of a set image hologram is the dove stamped on all VisaTM credit cards.

When you are using holographic foil with set images, the press must be equipped with a hologram registration unit. Hologram foil with set images, unlike regular or random pattern hologram foil, must be registered to the press. Each image on the foil has a registration mark. A hologram registration unit, by means of an electric eye, reads each registration mark

The Worldwide Authority on Holographic Machinery

When looking for Holographic Origination and Reproduction equipment, there is only one authority — James River Products. Since 1985, James River Products has been the world leader in holographic machinery production. Their experienced, dedicated staff has used its eye for innovation to create the highest quality product line available, ranging the whole scope of holographic needs. Products include:

- **Shutter Controls**
- **Photo Resist Plate Spinners**
- **Photo Resist Plate Dip Coaters**
- **Mechanical Recombining Systems**
- **Complete Electroform Facilities**
- **Lab Embossing Machines**
- **Precision Narrow Web Embossers**
- **Wide Web Embossers**
- **Narrow and Wide Web Coaters**
- **Rotary Die Cutters**
- **Reciprocating Die Cutters**
- **Sheeters**
- **Complete Holographic Origination Labs**
- **Electronic Machine Controls**

To satisfy your holographic machinery needs, look no further than James River Products.

Call, write or fax for more details or to arrange a demonstration:
James River Products, Inc.
800 Research Road
Richmond, Virginia 23236 USA
Phone: 804-378-1800
Fax: 804-378-5400
Telex: 903117 JRP-RCH-VA

and thus assures the image on the foil is centered over the die in the press. When the press cycles, the image on the foil is transferred to the stock by means of the die, heat, and pressure. A press without a hologram registration unit mounted to it cannot hot- stamp set image hologram foil.

Finally, one needs a source of foil. Today most major foil manufacturers produce hologram foil. Of course, price will vary depending on quantity needed, the size of the image, and its quality. It is in the area of foil that the greatest reduction in cost has been found. By vertically integrating the produc- tion process of origination, going to film and produc- ing the foil, foil companies save the cost of middlemen and shipping.

Holograms enjoy wide use in today's market. In the USA and throughout the world, holograms are used successfully. Manufacturers have realized for a long time that consumers are attracted to the perceived added value which hot-stamping adds to labels and cartons. Holograms take this concept one step further and are particularly successful when displayed at point of purchase.

Added to a hologram's marketability is its inherent security. With the advancement of copier machines and the market for counterfeit products, holograms offer some of the highest quality and easi- est ways to produce assurances of security (ED: please note that last year, however, millions of counterfeit holograms were seized from a company involved in producing bogus software--so be carefulf). It is expected that the field of security holograms will soon surpass that of general marketing.

Producing a finished piece with a hologram on it has never been easier. The actual process of setting up a press to stamp holographic foil is as easy as set- ting up a fiat stamping job." (Source: a speech Brandtjen gave at a foil stampers and embossers' conference, March 1993)

To give you an idea of the total costs for embossing, and then hot-stamping, here is a report on a magazine cover done in the last two years:

A Typical Magazine cover

A issue of a trade publication for magazine purchasing agents was devoted to embossed holography and its costs. The editors commissioned a hologram and had it hot-stamped on the cover of the magazine. They reported the following costs involved:

- 3-D models cost in the range of $1,000.00 to $4,000.00 depending on the amount of detail (see section on model-making for complete details on this process).

- Masters cost between $2,000.00 and $5,000.00 depending on the complexity.

- The cover of the magazine had a 7" x 5" hot-stamped hologram of a 3D model. Modeling and mastering, they reported, cost about $5,500.00. Embossing the hot-stamp foil cost approximately $0.10 per image. hot-stamping cost $0.03 to $0.06 per image.

- Embossed foil cost $7.20 for 1,000 inches. Non- embossed foil sold to embossers cost $1.46 for 1,000 inches.

- Polaroid has a proprietary process and photopolymer material which it calls "Mirage holograms". On aver- age it costs $0.10 per square inch for production runs and the price can drop to $0.05 per square inch with volume. The mastering would cost $8,000.00 to $15,000.00. The holograms can be produced from.66 x.75 inches to 10 x 14.5 inches.

In general, the magazine editors found that the big appeal of embossed holography is that it definitely increases the percentage of returns on direct mail pieces and increases circulation for magazines and books. On the down side, they noted that the mastering cost is the one item that keeps this from being very cost- effective in short runs.

Prices going down

Since the above magazine was done some two years ago, prices have been coming down. Don Tomkins, who works for Astor Universal, a foil supplier in Santa Fe Springs, CA, spoke about the costs involved in mass- producing holograms on foil.

A large difference in pricing can be made right away between stock pattern foil and custom images, said Tomkins. He estimated that the average price for stock images among most suppliers is approximately $2.50 for a 1" x 200' roll, whereas that cost could be between $4.00 and $10.00 for the same amount of foil embossed with a custom image. This does not take into account the separate origination fees (modeling, mastering) which could add additional thousands to the production of custom images.

There are a variety of grades of foil depending on the substrate to which the hologram will eventually be applied: plastic, paper, and varnish board all have slightly different requirements. Jim Waitts of Crown Roll Leaf, a large foil manufacturer, points out that foil for security applications must have increased scuff resistance and other features to ensure the nondestructability of the image.

Some good news here is that foil costs have dropped remarkably in the last couple of years. Tomkins says this is a combination of better production efforts and market forces--there are simply more large foil makers trying to get into embossed holography.

Tomkins points out that, naturally, costs depend upon production volume. He also notes that some printers are set up to run relatively small production jobs, whereas others are set up with machines that specialize in millions of impressions.

Embossed Holography Step-by-step

Embossed holograms are created in several steps. We will present an overview of the steps involved.

Overview:

The most common steps, once the model (not a simple step) is made, are as follows:

1) The hologram master is made.

2) A rainbow transmission hologram is made on photoresist (photosensitive emulsion) from the master.

3) The photoresist is etched to relieve the hologram patterns.

4) The photoresist is plated with silver and nickel. The photoresist now behaves like a metal mold (information on coating photoresist plates for holography is included in our chapter on recording materials).

5) The metal mold, or shim, is removed from photoresist and now has holographic patterns on it.

6) This shim is used as a stamping die and stamps the holographic patterns into plastic.

7) The plastic with the holographic pattern stamped in it has a mirror-like backing so light comes through the plastic, strikes the mirror-like backing and, reflecting back out, displays the white light rainbow transmission hologram.

Let's look at these steps in a little more detail and then discuss the model making.

Steps 1-2: There is a possible shortcut here. Sometimes flat art can be directly exposed onto the photoresist material. This is recommended for simple projects only.

Photoresist is a very tricky medium to record on holographically. It is nowhere near silver halide in responsiveness and requires long exposures even with lasers of several watts of power. A small holography studio would have to be equipped with an expensive laser and heat-resistant optics. New alternatives to this medium of photoresist are being researched.

The typical turn-around time is four to six weeks to receive excellent photoresist plates that are fit for metallizing. The holographer should make several photoresist plates of good quality to cover any problems that might occur in the metallizing phase.

Steps 3-4: After the photoresist hologram is checked for clarity, brightness and overall quality it goes to the metallizing stage of production. A thin layer of silver is deposited on the photoresist. Silver by itself cannot withstand the stamping pressure, so additional coats of a nickel-based material are deposited to reinforce the back of the silver. When it achieves the desired thickness, the nickel-silver shim is pulled from the photoresist plate and this becomes a metal mother die.

Steps 5-7: The first shim must be perfect. This first shim can have several shims made from it and in turn several shims made from those. The heat and pressure of embossing thousands of holograms wears out the shims so extras should be made. Any deterioration becomes very obvious if the shims are not changed regularly.

Producing the final hologram is the job of the embosser. Embossed holograms are made by stamping the shim onto aluminum. Heated polyester material which has a metallized backing is used less often. Aluminum is softer and less destructive of the nickel master. Although not used widely, colored metallized backing is an option. Most often the silver color is selected.

Embossing Pulsed Holograms and Holographic Stereograms

Pulsed hologram images are used in embossing and have been very successful. Be sure the pulsed holographer knows that you want this to be an embossed hologram. It is a more complicated approach which may include a step to reduce the pulsed hologram to the six-inch square size of the embossing machine, or whatever limitations your embosser has. Do not use this approach if your turn-around schedule is too tight. Allow enough time to make sure all steps take place with breathing room if a problem occurs. The results of pulsed embossing are stunning and it is worth pursuing.

Preparing Artwork For Embossed Holograms.

Embossed holograms made for advertising or artistic effects have a common problem. Most of them are intended to be seen under any available lighting condition, including very harsh light like fluorescent. This means that the hologram you make cannot have a great deal of projection or depth, because fluorescent light does not display depth well. This means that 3D objects, if they are used, must be very shallow. The most common objects used are either flat artwork or very thin, miniature models.

3D Holograms: A 3D model is frequently made to act as subject in 3D holograms although very lifelike images can be created with photos. Close consultation with the holographer before and during the making of the model is important (more detailed information about this subject is contained in the chapter on model-making in the beginning of this book).

2D / 3D Holograms: A 2D/3D hologram is a hologram made up entirely from flat artwork appearing on two or more levels. If the flat artwork is entirely one level, the hologram is called a "2D" hologram. If it is layered on two or more levels, the hologram is called a "2D/3D" hologram. The attraction of a 3D holo- gram is that you have a model you can see before the hologram is made and, although shallow, there is some 3D depth to the final embossed hologram. Although one might think that a 3D hologram is the "best money can buy", 2D and 2D/3D holograms have several advan- tages.

Advantages of 2D/3D holograms from the holographer's point of view:

• 2D/3D holograms are all made using exactly the same holography setup, which never has to be changed.

• The "object" for a 2D3D hologram is simply flat artwork, so there are no object motion problems or object mounting challenges.

• All of the image information for a 2D hologram or the 2D part of a 2D3D hologram is contained on the surface of the hologram, so the 2D part of the image can be made much brighter than an ordinary 3D hologram without loss of image detail.

Advantages of 2D3D holograms from the point of view of the graphic artist and the ultimate user:

• Graphic artists are well acquainted with two-dimensional media, so their techniques and skills are easily adapted to designing 2D and 2D/3D holograms.

• It is possible to make mock-ups of 2D/3D holograms as an aid to making a sale, without going through the hologram mastering process and expense.

• The graphic artist who designs the 2D or 3D holo- gram can define all aspects of the image: artwork, imagery, colors, and image placement. The hologra- pher may assist the graphic designer, however, in proper color choices because the former sometimes possesses a better understanding of spectral events, and can therefore produce exciting kinetic effects.

• 2D/3D holograms can be viewed in any kind of light- ing without difficulty, making them suited for the many applications in which there is little opportu- nity to control the lighting, such as packaging, point-of-pur- chase displays, advertising, anti-counterfeiting, tex- tiles, clothing, toys, greeting cards, and decorations.

• The greater brightness and sharpness of 2D/3D holo- grams compared to other types of holograms make them excellent eye-catchers.

2D and 2D /3D Artwork: The artwork for 2D and 2D/3D holograms is essentially the same sort of flat art- work that any ordinary printer would use. The only dif- ference is that, since a hologram allows a person to look around a foreground image, the background should be complete, without "cutouts" for the foreground. color separations are usually prepared from the line art, and the color for each separation is specified. It is also possible to produce 2D/3D holograms from color photographs, though it is more complicated.

Line art for 2D and 2D/3D holograms should have an unbroken line or boundary for each region of any given color. Each layer in the image should have its own separate artwork. Color overlays should be provided with line art to designate the colors of the various parts of the image in each layer. Each layer should have its own color overlay.

SECTION 4

Equipment

11

Lasers

Since the laser is at the heart of making holograms, a basic understanding of laser principles and operation is essential for creative applications as well as successful buying. In this chapter, we will firs race some of the important developments that led to the modern laser, consider what lasers do and how they work, and discuss the features of lasers important to holography.

A Brief Historical Background

Today, the laser's widespread applications range from playing audio compact discs to pinpointing targets for bombs. Lasers have become indispensable tools for such diverse fields as construction, pollution detection, medical surgery and of course, holography.

Few people realize that this technological triumph started as a rather esoteric paper written by Albert Einstein in 1917. In his paper, Einstein claimed that atoms could generate light and other forms of radiation by a mechanism he called stimulated emission. Previously it was thought that atoms and molecules could generate light only by a mechanism called *spontaneous emission*.

To understand Einstein's distinction we need to define two important ideas that have shaped much of modern physics, the idea of the photon and the idea of distinct energy levels in the atom. The photon is an irreducible unit of light energy. Light beams are thought to be composed of a finite number of photons. The theory of the photon shattered the belief that light could be infinitely subdivided into smaller and smaller units of energy. The theory of distinct energy levels in the atom, or quantum levels, upset a similar belief, the belief that any amount of energy could be injected into an atom. In fact, as quantum theory

suggests, the energy of atoms changes in "quantum jumps" from one energy level to another. An atom cannot have an amount of energy between two levels.

If we put these two ideas side by side, we reach an interesting conclusion: photons that travel past atoms may be absorbed by the atom, only if the photon energy equals the transition energy needed to excite the atom to a higher quantum level. If a photon passes an atom with one half the energy needed to put the atom to the next energy level, the photon cannot be absorbed by the atom. If the energies are the same, absorption will occur. Conversely, if an atom moves from a higher to a lower energy level, then a photon will be emitted. The energy of this photon will equal the atom's transition energy.

In nature, atoms are frequently excited to energy levels above their lowest state, ground state. However, in a tiny fraction of a second these atoms emit a photon and return to their ground states. This emission occurs without any outside influence and is therefore called spontaneous emission.

Einstein suggested that light might also be produced by a process he called stimulated emission. Previously, it was thought that if a photon passed by an excited atom, the photon might be absorbed if its energy matched the atom's transition energy or the photon simply would pass the atom. Stimulated emission was Einstein's alternative to this thinking: the photon passing by the excited atom would "stimulate" the emission of a photon from the atom. Thus two photons would leave the atom simultaneously.

The main point of Einstein's argument was that two processes, not one, explain the generation of light in nature.

But this theoretical point raised a technological challenge: building a source that would generate radiation such as visible light primarily by the mechanism of stimulated emission rather than spontaneous emission. The problem was in establishing a large number of excited atoms, since atoms tend to remain in their ground state or return very quickly if they are excited.

The first success came in 1953 when Charles H. Townes created the maser (an acronym for Microwave Amplification by the Stimulated Emission of Radiation). Townes found a way to separate excited ammonia atoms from ammonia atoms in the ground state. The maser generated microwaves, a form of radiation less energetic than light, by stimulated emission.

The difficulties of generating visible light were only overcome about seven years later when Theodore Maiman made his first laser (an acronym for Light Amplification by the Stimulated Emission of Radiation) by putting mirrors on the two ends of a synthetic ruby and illuminating it with a powerful flash lamp. Within a few years, Emmett Leith and Juris Upatienks made the first laser three-dimensional holograms.

Laser research over the past 30 years has multiplied the different types of lasers available and their range of features. Most lasers fall into three classes:

(1) solid state lasers, like Maiman's ruby laser,

(2) semiconductor lasers, which are typically found in CD players

(3) gas lasers which are the type commonly used in holography. This classification excludes some of the more exotic lasers like the chemical dye lasers.

Lasers also generate electromagnetic waves of many different types, one of which is visible light (actually, the laser by original definition produces only visible light, but since laser principles are used in other devices like the maser, we will loosely call all these devices "lasers")

Electromagnetic waves generated by lasers are conceptualized as "ripples" in the electromagnetic field that permeates all of space. These waves are characterized by their wavelength, which is the distance between successive wave crests, and by their amplitude, which is half the height between a wave crest and a wave trough.

The wavelengths generated by lasers range from long wavelength microwaves to short wavelength X-rays. Laser power, which is proportional to the amplitude squared, can vary from a fraction of a milliwatt to trillions of watts.

Common to this diverse array of lasers are a few basic principles which we consider next.

II. How a Laser Works

The road between Einstein's paper on stimulated emission and Theodore Maiman's ruby laser was filled with technical obstacles. The major problem was in energizing the atoms so that a majority were in an excited state. Remember that stimulated emission will occur only when a photon passes an excited atom and the excited atom's transition energy from excited state to ground state equals the photon's energy.

The problem is that excited atoms decay to their ground states in extremely short times. The net result is that a fraction of atoms are excited at any given time, leaving little chance for stimulated emission to occur. Only if there is a population inversion, or a pre- dominance of excited atoms with transition energies matching the photon energy, will laser action occur.

To see how population inversion is achieved, let us look at the Helium-Neon laser, the laser most commonly used by beginning holographers. Figure 11.1 illustrates three of the important energy levels in neon.

If a neon atom is excited to the second energy level, it will decay quickly back to the 1st energy level. Atoms excited to the third energy level decay slowly back to the second and are therefore called metastable. Remember that each time an atom decays to a lower energy level, a photon is emitted. For example, the decays from the third to the second energy level emit photons with a wavelength of 632 nanometers (632 billionths of a meter), which is the characteristic

No More Mode-Hops

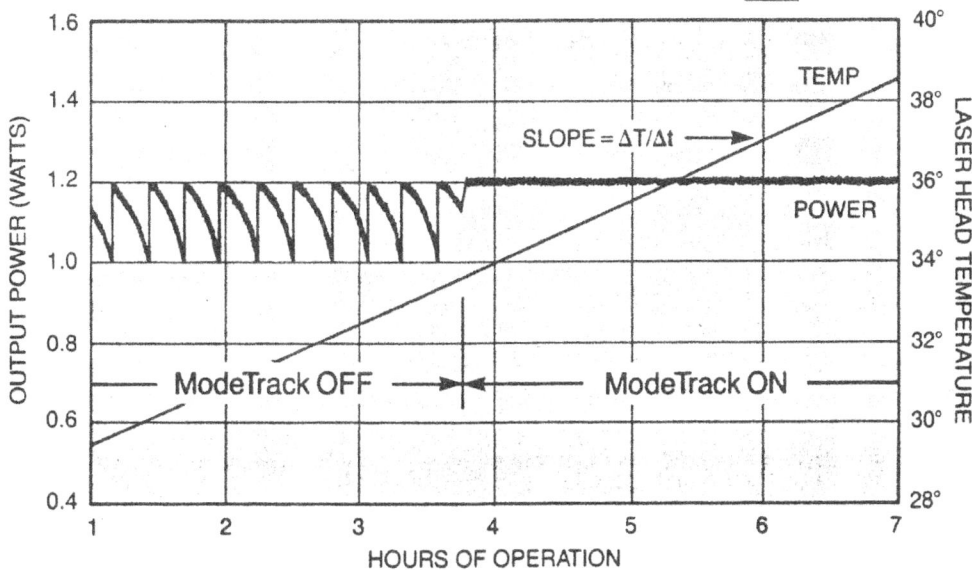

Graph: OUTPUT POWER (WATTS) vs HOURS OF OPERATION, with LASER HEAD TEMPERATURE on right axis. Labels: TEMP, SLOPE = ΔT/Δt →, POWER, ModeTrack OFF ←→ ModeTrack ON.

Stable Operation Guaranteed

The Innova® 300 Ion Laser System with ModeTrack™ eliminates mode-hops. It's that simple. Now you can make long exposure holograms on the first attempt. And you'll experience a lower reject rate in your high volume holographic applications.

Changes in ambient air or cooling water temperature are the main causes of mode-hops in ion lasers. ModeTrack utilizes the Innova 300's system CPU to continuously monitor the single-frequency stability and controls the etalon to prevent mode-hop.

The Innova 300 also features PowerTrack,™ an actively stabilized optical cavity. Power-Track automatically provides both optimum output power and unmatched long-term stability. PowerTrack coupled with ModeTrack provides you with hours of hands-off operation, without a single mode-hop.

No other laser offers either of these capabilities. No other laser can increase your productivity and quality like the Innova 300.

Innova 300. The Laser for Holography.

To learn more about Innova 300 ModeTrack, please call 1-800-527-3786 or 408-764-4323; fax (800) 362-1170 or (408) 988-6838.

Benelux (079) 621313 • France (01) 6985 5145 • Germany (06074) 9140 • Japan (03) 3639 9811 • United Kingdom (0223) 424065

✳ COHERENT®

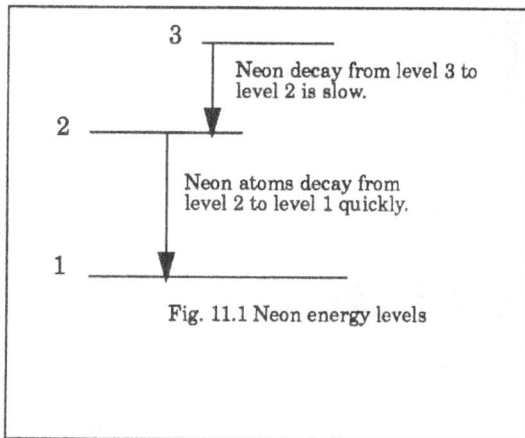

Fig. 11.1 Neon energy levels

red light of the He-Ne laser.

This configuration of energy levels suggests a promising possibility: if neon atoms can be excited to the third energy level, then their slow decay to the second energy level together with the rapid decay of atoms in the second energy level to the first or ground level will result in a predominance of atoms in the third level over atoms in the second level. In other words, population inversion occurs between the second and third level, resulting in stimulated emission.

The trick to exciting neon atoms to their third level is illustrated in figure 11.2. It turns out that it takes the same amount of energy to excite helium to its second level as it does to excite neon to the third or metastable level. When excited helium atoms collide with ground state neon atoms, the energy is transferred, exciting the neons to the third energy level. Once neon atoms are excited to the third energy level, population inversion between the third and second occurs.

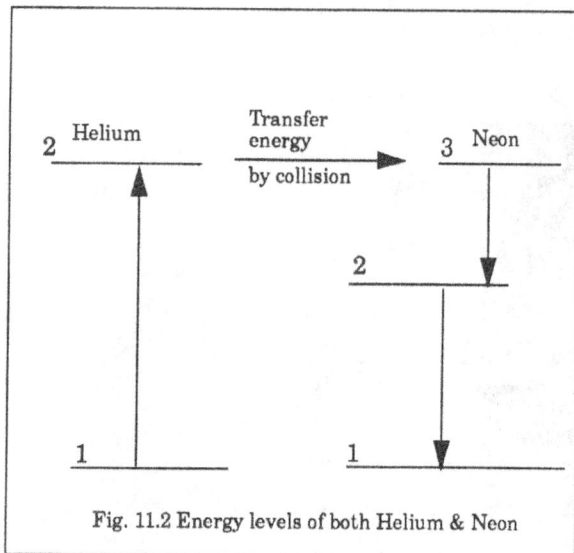

Fig. 11.2 Energy levels of both Helium & Neon

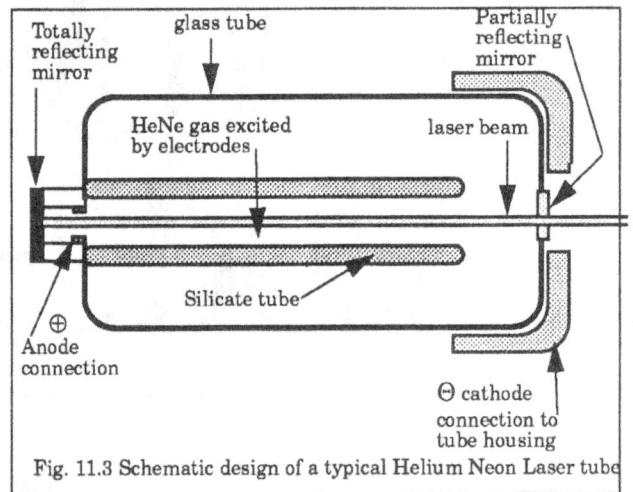

Fig. 11.3 Schematic design of a typical Helium Neon Laser tube

Figure 11.3 schematically illustrates a typical HeNe laser tube. While the entire glass tube is filled with low- pressure helium-neon gas, only the central region near the axis of the tube is excited. The excitation is accomplished by two electrodes that send high-speed electrons through the gas. An anode (positive charge plate) is attached to the left end of the tube and the cathode (negative charge plate) is attached to the tube housing. Charged particles traveling between the anode and the cathode collide with, and thereby excite, helium atoms mostly in the silicate tube region. These excited helium atoms then collide with neon atoms and create a population inversion between Level 3 and Level 2 neon atoms.

Decays from the third to the second level set off a chain reaction of stimulated emissions. The cascade of photons is directed through the narrow silicate tube toward the mirrors at the opposite ends of the laser. Most of the light bounces back and forth between the mirrors in order to induce more stimulated emissions. This increases the number of photons in the light beam. A small fraction of this beam is allowed to escape through the partially-reflecting mirror on the right side of the tube. This light that leaks out is the characteristic red beam of light generated by a laser.

Now that we have explored basic laser operation, we can turn to the laser features important in holography.

Laser Applications in Holography

The three common lasers used in holography are the Helium-Neon lasers, Krypton lasers and Argon lasers. All three are gas-filled lasers. Unless you are not con- strained by cost, the best starter laser is the Helium- Neon (He-Ne) laser. The major difference among the three kinds is power.

Krypton and Argon lasers typically generate 2-4 W, thousands of times more power than the weaker HeNe lasers. This intensity will allow you to make larger holograms and work with a greater variety of photographic film or emulsions. Expect costs to start in the :ens of thousands of dollars.

The best starter laser is the He-Ne laser, which is reliable and has very long lifetime. A bottom-of-the-line model sells for as little as $200-$300. This starter model will probably have a power rating of 0.5 to 1 mW - about 100,000 times less intense than an average light bulb. You will have no problem seeing characteristic red dot if you shine the laser on a wall, but creating holograms may be difficult because of the limited light intensity. For a few hundred dollars more, a laser in the 2 to 5 mW range can be purchased. Serious holographers may want to purchase HeNe lasers as powerful as 50 mW. Although holograms will be easier to make, you will still be limited by the photographic film or emulsion you can use. Most HeNe holograms will require a silver halide emulsion.

To conserve costs, you may consider purchasing a sed laser, perhaps reconditioning some old tubes, or building your own power supply.

Besides power, other important laser features are continuous/pulsed emission and coherence. The helium neon laser just described operates by continuous emission. The light emitted by the laser is a steady beam of photons that is generated by the continuous build-up of Level 3 electrons. Another mode of operation is pulsed emission. If the atoms in a laser are excited very suddenly and then decay all at once, the result will be a short flash (pulse) of laser light. This type of laser is more expensive.

An example of a pulsed laser is the Q-switched laser. A typical Q-switch basically controls whether the cavity mirrors in the laser will absorb or transmit light (the mirror switches from a high "Q" or "quality" of transmission to a low quality of transmission). While the mirrors are in the low Q mode, they absorb light and prevent a cascade of stimulated emission by keep- ing the light waves moving between the two mirrors at a low intensity. This gives the atoms ample time to build a large population inversion. When the mirror switches back to a transmitting mode (high Q), a cascade of stimulated emissions produces a short, intense burst of light.

Holograms can be made either with pulse or continuous lasers. Pulsed lasers have the advantage of freezing motion like a high-speed camera shutter. The exposure time is merely the time it takes for the light pulse to sweep across the object. Continuous lasers have lower light intensities that usually require a longer exposure time for the image to form. There- fore high-speed holography is not possible with a continuous laser system. Furthermore, the holography must be done on a vibration-free table because any motion would blur the hologram. Most holo- grams are made with continuous rather than pulsed lasers simply because they are cheaper.

For both continuous and pulsed lasers, coherence is the feature most important in creating a hologram. Two waves are coherent if they move in phase with each other, meaning the crests and the troughs of the two waves continue to line up as the waves move. Laser light is coherent because the waves generated by all the different atoms are in phase or very nearly in phase. This is one of the basic properties of stimulated emission.

Coherence is what makes three-dimensional holography possible. Ordinary two-dimensional pictures merely capture light intensity. This intensity is represented by the relative brightness and darkness on a photograph. Holograms actually capture the interference pattern between the light from the object and a reference beam of light.   T h u s when the hologram is developed, instead of having an actual image, the photographic plate has small microscopic interference fringes. When a reference beam is shone through these fringes, the light diffracts. The resulting light that reaches our eyes in an exact replica of the light that bounced off the object that was photographed. This creates the illusion that we can see the object from different angles.

There are two types of coherence holographers need to understand: temporal coherence and spatial coherence. To understand temporal coherence, imagine that we split a laser beam into two halves and let each half run side by side separated by a divider. If the waves are perfectly coherent, we can tell from one side of the divider what the phase was on the other side, next to us as well as very far up the beam path. However, with an imperfect coherence, we can only predict the phase a limited the distance up the beam path. Beyond that point, the phases of the two beams do not correspond in any predictable way. This maximum distance is called the coherence length.

Thus temporal coherence or coherence time is simply the coherence length divided by the speed of light. When making a hologram, it is important that the dimensions of the object are less than the coherence length. A good rule of thumb is that the laser's coherence length should be about two or more times the length of the object. This is because holograms are constructed by dividing the beam of light into two beams, an object beam and a reference beam. The object beam reflects off the object onto the photographic plate while the reference beam simply illuminates the photographic plate. When the two beams rejoin at the photographic plate, they must still have a predictable phase correspondence in order to record phases from the different points of the object (remember, it is by recording this information that holograms achieve their three-dimensional effect).

If the dimensions of the object are greater than the coherence length, then those parts of the object beam that out-travel the reference beam by more than a coherence length will have no way of comparing their phases with the reference beam when they reach the photographic plate. By losing phase information, the three-dimensional effect will be lost.

An inexpensive Helium-Neon laser typically has a coherence length of about 10 cm, sufficient for holographing small objects. If you want to make holograms of larger objects, make sure to find out the coherence length before you make your purchase.

Another type of coherence is spatial coherence. One way of testing the spatial coherence of your laser is to shine it on a wall. If you are fortunate, the pattern will be a bright dot that gradually fades towards the edges. When this occurs, a laser is operating in the TEM_{oo} mode. This means that there are no interference patterns across the beam caused by stray transverse electro-magnetic (TEM) light waves. The modes refer to the way the light moves at angles to the beam line inside the laser cavity. Certain mixtures of straight moving and angling beams (represented by the two subscripted numbers after "TEM") create characteristic patterns. If the pat- tern were a bright ring with a dark hole in the middle it would be operating in the $TEM_{(10)}$ mode. For a laser to create holograms, it is essential for it to operate in the TEM_{oo} mode. Otherwise the object will not be illuminated with an even phase front, and the hologram will be distorted.

Finally, if you have decided to purchase a laser, you may wish to contact any of the manufacturers of lasers in our listings.

The above article was contributed by:

Gabe Paulson
Berkeley, California

Ion Laser Guide

Whether you are buying your first ion laser system and are unfamiliar with the different choices available to you, or are planning to acquire an additional system and need to be up-to-date on the latest technology, there are a number of different criteria that you will want to review before you make your purchase decision.

You application will have specific requirements which only you can define. These include such things as wavelength coverage, output power, optical noise, day- to-day reliability, plasma tube lifetimes, and long-term stability. Additionally, your facilities may impose limitations in terms of available electrical power, cooling water, or air conditioning. Finally, you will need to review post-installation supplier support in terms of onsite service policies, spare parts availability, telephone technical support, and commitment to your productivity.

Once you have defined your needs in these areas, it will be much easier to determine which model of ion laser and which manufacturer best meet your require- ments.

Ensure that the facility and the laser you select are compatible, and review in advance the support capabilities of the manufacturer(s) you are considering.

Ion Laser Basics:

Ion lasers operate primarily in one of three configurations: multiline, single-line, and single-frequency. Many systems are capable of operating in each of these configurations through the addition of optical components. There are also "application-specific" laser systems which operate in a single configuration and provide advantages for a given application.

Multiline and Single-line:

In the multiline configuration, the optical cavity consists of a high reflector and a partially transmissive output coupler. The output will consist of some number of discrete wavelengths (or lines) which lase simultaneously. The available wavelengths range from UV to IR (Ultraviolet to Infrared) and will be determined by the lasing medium (argon, krypton, mixed-gas, etc.), discharge current, and the coating of the optics. In an argon laser, for example, the multiline output will usually consist of eight lines from 457.9 nm to 514.5 nm. This output is suitable for many applications such as tunable laser pumping- -that are not wavelength specific.

Single Line

There are other uses for the ion laser, such as spectroscopic applications, that require a single, narrower

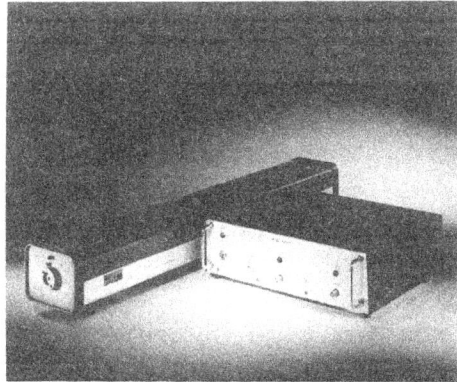

excitation wavelength. These applications require the ion laser to operate at only one of the available wave lengths. This is accomplished by inserting a Brewster prism into the cavity of the ion laser. This prism introduces dispersion between the available lines, and by angle tuning allows only a single line to lase.

In this configuration the user can easily change the angle of the prism and high reflector to select any of the available wavelengths. The line width of the laser output in the single-line configuration is approximately 6 GHz.

Single-Frequency
Applications such as holography require a much narrower line width than is available in the single-line configuration. The narrower line width is achieved by inserting an additional optical element, the etalon, into the laser cavity to achieve true single-frequency opera- tion.

The insertion of an etalon into the ion laser cavity at a small angle will allow it to act as a frequency filter and suppress all but one single longitudinal mode. In this configuration, the instantaneous line width of the laser is reduced to a few MHz. This will broaden, due to environmental influences, and the actual bandwidth will depend on factors such as ambient temperature, stability, and time of exposure.

Sealed-Mirror
Sealed-mirror designs offer the user some significant advantages. Higher powers are available, maintenance is reduced and plasma tube lifetimes are extended. At the same time, there is usually a savings realized by the customer, both in terms of the initial purchase price, and in the subsequent cost of utilities and replacement parts.

Removing the Brewster window results in an increase in power, improvement in beam quality, and enhanced plasma tube lifetimes. Additionally, since there is no optic/window cleaning required in this area, the risk of operator-induced damage is eliminated.

Output Power
The output power from an ion laser varies from tens of milliwatts to 25 watts and is a function of amazing medium, discharge current, and optical con- figuration. Typically, ion lasers are referred to as either "Small frame" or "Large frame". The small frame ion laser will typically operate from 208 vac, three-phase electrical service and has a cavity length of 1 meter. The large frame ion laser requires 480 vac, three-phase electrical service and has a cavity length of about 2 meters. Output powers differ significantly between small and large frame systems.

Recently, with the introduction of application-specific ion lasers, performance in some areas has been improved. For example, as a result of a new hard-seal technology, small frame lasers--traditionally limited to 6 w multiline visible or 500 mW multiline UV--now provide up to 10 w multiline visible or 1 w multiline UV output.

Operating Characteristics & Features

When you review the brochures describing ion laser systems, you will find that there are different operating characteristics and features which may be unfamiliar to you. The following terms may be useful in providing you with a generic description of these items.

current regulation: The plasma tube discharge current is held constant. Output power fluctuations may occur due to changes in the resonator alignment, optics becoming dirty, etc.

light regulation: The output power of the laser is held constant by monitoring the out put power level on a detector and changing the plasma tube discharge current to compensate for changes in output power.

actively stabilized cavity (*not to be confused with light regulation): One cavity mirror is mounted in such a manner that it may be angularly adjusted rapidly in the vertical and horizontal axes. This allows the laser to be optimized for output power, mode, and noise, or to maintain constant alignment for fixed beam position.

power stability: A measure of peak-to-peak low frequency power fluctuations over a given time period.

optical noise: Typically is measured with a 10Hz-2MHz photo diode driving a resistive load. Values are usually expressed as an rms value for the 514.5 nm line in argon lasers and the 647.1 nm line in krypton lasers.

beam pointing stability: There are two components to beam pointing stability: angular deflection and position offset. Direction of deflection refers to changes in the angular propagation of the output beam. Position offset refers to translation of the beam at the output coupler.

warm-up time: Describes the amount of time necessary for the ion laser to reach specified output power and power stability.

diagnostics: Built-in troubleshooting aids. These vary widely from system to system and can be as simple as a few indicator lights or as complete as onboard CPU's with operating parameters displayed in English on an LCD display.

preset power levels: in many applications it is desir- able to rapidly switch back and forth between two or more power levels. Some systems have memory capabilities and will allow the user to set the desired levels and switch quickly between them by pushing a button.

temperature stabilized etalon: To provide a more stable single-frequency output, the etalon is enclosed in an oven, raised to a temperature well above ambient, and the temperature held constant.

Actively stabilized etalon: Used in conjunction with the temperature-stabilized etalon. Single-frequency output is monitored and a feedback loop is used to change the temperature of the etalon. This system continuously matches the etalon transmission peak to the resonant cavity mode. Power losses associated with frequency drift and the instantaneous frequency shift associated with mode hops are eliminated.

Productivity

The proper selection of an ion laser should allow you to increase the quality and quantity of your work. As discussed above, there are many features available on ion lasers today which help in this area. Active cavity stabilization reduces warm-up time from one-two hours to one minute. This means that you can get more done each day. Single frequency applications benefit significantly from an actively stabilized etalon. Not only are mode-hops eliminated, but warm-up time in these applications are reduced 50-75%.

When discussing ion lasers with your supplier, it is worthwhile to spend a little time specifically reviewing the factors which will allow you to be most productive in your work.

Versatility

In making a decision on the ion laser you acquire today, make sure that you consider future applications as well. Many of the ion lasers available today can be upgraded in the future to offer broader wavelength coverage or higher output powers. Ass an example, low power (4 watts) small from blue-green argon laser from some manufacturers can be easily converted to red, violet or UV krypton versions or upgraded to higher power (10 watts visible or 1 watt UV) argon configuration.

Upgradability provides a cost-effective way of managing your budget as your application needs change.

Facility requirements

Ion lasers are not efficient. A small frame laser with a 6-watt multiline output will require an input power from the wall of about 15kW. As a result there is a significant dissipation of heat, and water cooling is required. Typical facility requirements are as follows.

Table 1: Facility Requirements for Lasers

Parameter	Small Frame	Large Frame
Electrical from wall	220 vac, 3-phase 40-60 amps	480 vac, 3-phase, 60-70 amps
Heat dissipation (wall input)	10-1-5 kW	35-50 kW
Cooling water flow	9.5 liters/minute (2.5 gpm)	22.7 liters / minute (6 gpm)
pressure	2.00-4.23 kg/cm2 (25-60 psi)	2.47-4.23 kg/cm2 (35-60 psi)
temperature	10-35 C (50-95 F)	10-35 C (50-95 F)

Additionally, the air-conditioning in your facility may have an impact on the stability of the laser beam, especially if you are planning to undertake single-frequency applications or have a complicated optical system. Generally, a well-designed facility will have air temperature control of +/- 1 C, no direct air flow on the laser or optical train, and have the laser/optical train vibration isolated from the environment.

In planning the total cost of acquiring your ion laser, it is a good idea to review the capabilities of your existing facilities to see if additional power or cooling water lines will need to be installed.

Post-Installation Support

There are many questions about support that you will wish to have answered:

What is available in the way of telephone technical support?

Is service performed at my facility or do I ship the laser back to the factory?

Where is the nearest field service engineer located?

How quickly can I get replacement parts? What are the warranty/guarantee provisions for the product? What are the service contract options?

This Ion laser Guide has been provided by:

Coherent, Inc.
Laser Group
5100 Patrick Henry Drive
Santa Clara, CA 95054
1-408-764-4323

WHICH LASER SHALL I BUY?

This is one of the most commonly-asked questions when considering a holographic setup and also one of the most expensive.

This section will try to provide a guide to the lasers that are available for holography, with their associated recording materials, plus an indication of the direction of laser developments that are relevant to holography. Further, there will be some basic advice on obtaining the optimum output from your laser: tuning for maxi- mum output power and cleaning the optics.

The Helium-Neon (HeNe) laser
The ReNe has certainly dominated holography for many years. Its technology has been accepted by all holographers, from hobbyists, commercial to R&D establishments. The convenience of air-convection cool- ing, commonly available power source 110/240 VAC, reliability, low maintenance and reasonable pricing have led to this situation. There have been no significant developments in HeNe lasers up to about 35mW output during the last few years.

Since these lasers are low-gain devices, there is a limit to the maximum output power per unit length of active lasing medium of approximately 40 to 50mW per meter. The highest power available is about 90mW which is due to the tube manufacturing and physical constraints. One effect observed is that as the output power increases (increasing the tube length) the coherence length (overlap interval for interference) reduces. This coherence reduction can be improved by the use of an etalon in the laser cavity. This selective glass component allows a restricted range of laser occilations within the optical cavity. The negative side to this is that the output power will be typically reduced by about 50%. The best cur- rent compromise is an output power of about 50mW with a coherence length of approximately 8 meters.

The low maintenance costs of ReNe lasers, even if a replacement tube is required, is significantly less than that for an alternative red laser source, such as a krypton ion laser (647 nm). If the output power is adequate, then the change to a higher power ion laser should be resisted.

Silver halide recording materials, such as Agfa 8E75HD, sensitive to the red spectral region are readily available, plus there is a large amount of information on its use.

It is normal for holographers who require higher laser powers to consider argon ion (blue 459/488 + green 514.5nm) and krypton (red 647 nm) lasers. There have been few significant improvements in ion laser technology; reliability has improved, but the krypton laser still lags behind that of argon due to inherent physical difficulties. The output powers vary greatly from a few tens of milliwatts (air-cooled lasers) through to systems delivering output powers of about 10-watt per a single wavelength. Manu- fac- turers have produced external cavity devices to reduce instabilities and restrict modal changes. Again, the use of etalons can increase the coherence length from a few centimeters to several meters for use in large-format holography.

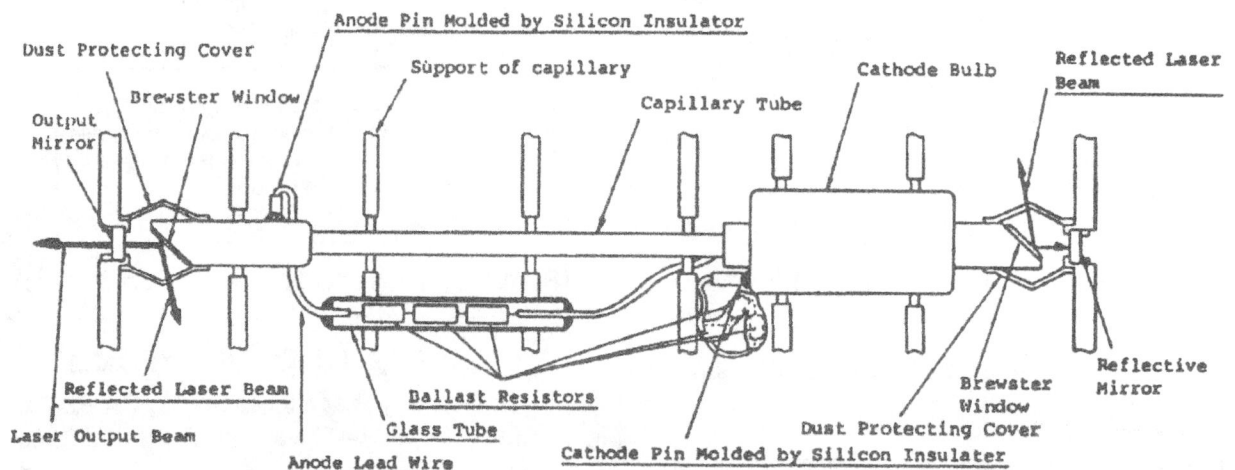

INTERNAL ARRANGEMENT OF A TYPICAL HE-NE LASER

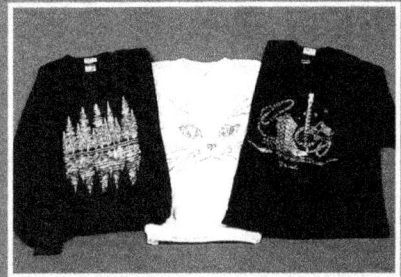

These ion lasers can be used not only with silver halide recording materials, but also with those that require the shorter wavelengths in the blue/green regions, such as dichromated gelatin (DCG; 459 to 514.5 nm) Photoresists (459 to 488 nm) and photopolymers ,6 (blue/green and red).

Recording materials such as photoresists, for producing embossed surface holograms, and DCG rely heavily upon lasers operating in the blue region. It is observed that a reduction in laser wavelength from 459nm to 442nm (Helium-Cadmium, or He-Cd laser) produces a tenfold increase in sensitivity.

The helium-cadmium (442nm) has become a very useful blue laser source with similar characteristics to those of the HeNe 2,7,8. Output powers up to about 100 mW are available but coherence length is typically restricted to 80 to 100mm. The latest development 2 is the production of a HeCd laser with over 100 mW output and extended coherence of between 600mm to 1000mm--these will be of great interest to large-format embossing system users. It may be possible to extend HeCd use to photopolymers by the use of fringe locking systems--any comments?

Visible semiconductor laser diodes (670 to 650nm) are in widespread use as a convenient replacement for low power (1 to 5mW) HeNe lasers (typical cost $100 to $200), but unfortunately they have only short coherence length, approximately 10 to 15mm. Further, the output beam is elliptical and requires additional beam shaping optics to obtain an output similar to that from a HeNe laser. There is a steady trend toward shorter wavelengths, although 630nm is the present commercial limit.

One of the most significant laser/holographic developments is in diode-pumped solid state lasers. Neodymium:YAG diode-pumped and frequency-doubled lasers have an output (532 nm) of up to 400mW with many meters of coherence. The system is pumped using diode lasers that are tuned to be efficiently absorbed by the Nd-ion at around 810nm. The power source is 110/240 VAC, with convection cooling. The systems are very compact, and reliable, but as yet expensive, typically $20,000 +. This will certainly prohibit their regular use for some time.

Pulsed Holography

Pulsed Holography is dominated by the Ruby laser. These systems remain fairly expensive and for specialist applications (portraiture, non-destructive testing, etc.). Output powers from 0.5J to 10 Joule are commonly available with pulse durations of about 20 nano-seconds.

It has been noticed that there is a steady interest in the use of pulsed frequency doubled YAG lasers 2 (532 nm) with output powers of approximately 0.5J to 1 Joule. The main advantage is the better reproduction of human skin texture--the green light does not penetrate below the skin surface to produce a wax-like form, as Ruby light does, unless a series of surface cosmetics are applied. Silver halide recording materials, such as Agfa 8E75HD and 8E56HD, respectively, can be used for these lasers.

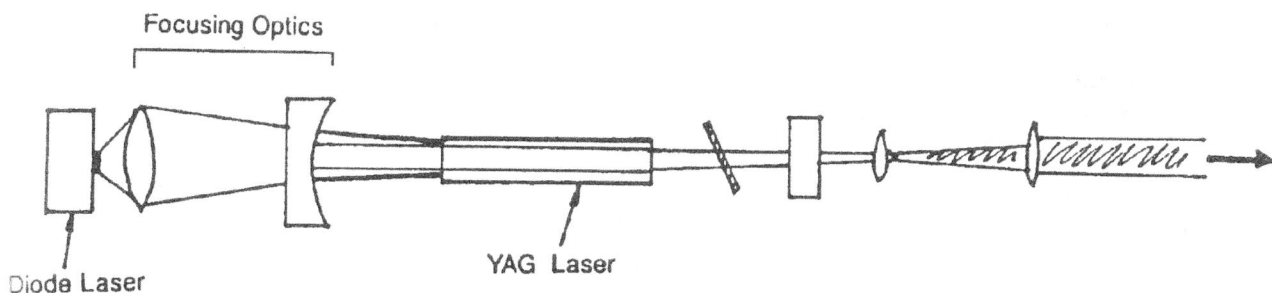

Focusing Optics

Diode Laser

YAG Laser

DIODE PUMPED FREQUENCY DOUBLED Nd: YAG LASER

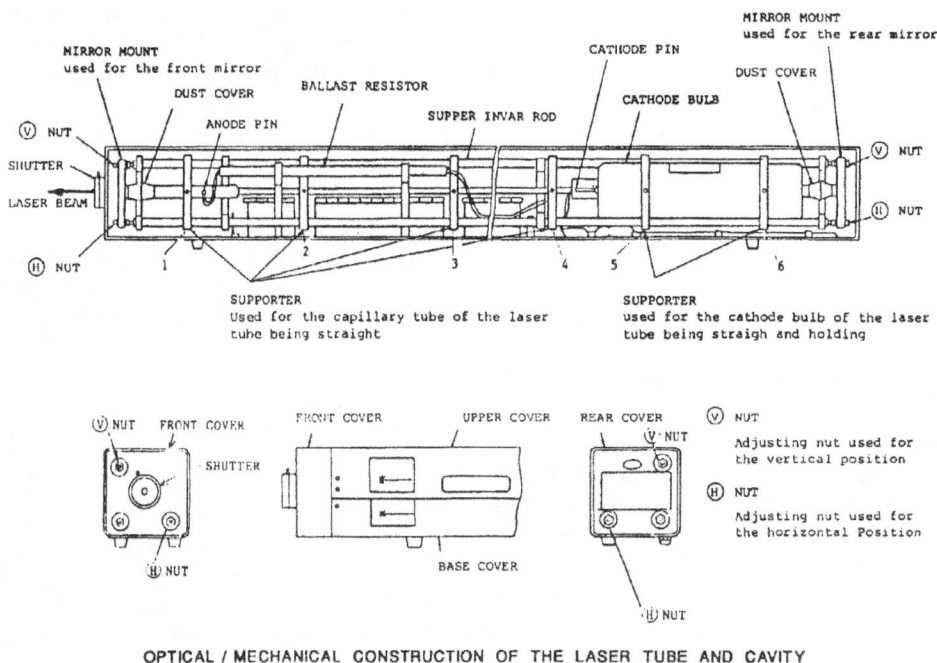

OPTICAL / MECHANICAL CONSTRUCTION OF THE LASER TUBE AND CAVITY

TUNING OF A LASER OPTICAL SYSTEM

(NOTE: The following is only applicable to lasers with mirrors that can be externally adjusted)

To improve the output power (when it has been observed to decrease) this is a basic tuning method that can be applied to most lasers. Ensure that the laser has reached its normal operating temperature/ characteristics--usually 30 to 60min after operation at maximum tube current. Place a suitable, reason- ably- quick-responding photodetector (e.g. a silicon photo- diode) connected to (preferably) an analog display (a digital display could be used, although the power out- put trend is easier to visualize with an analog device). Since the absolute output read- ing is only for record purposes, any suitable detector could be used.

Method: Move only one, say the top (vertical) ad- just- ing screw, at the rear of the laser (opposite end to the output port). This should be very slowly rotat-

ed in both clockwise and counterclockwise direction to increase the output power to a maximum. Next repeat this procedure with the horizontal control (diagonally opposite) and again adjust to maximum. Finally, return to the vertical control (top screw) and repeat.

If you suspect that the power of the laser is well below that which is expected, then the process of "walking in" of the laser should be performed. This involves moving both the output and rear mirrors progressively while maintaining the lasing output to optimize the lasing path through the active me- dium in the laser cavity (see later section for further detail).

For high gain lasers, such as argon and krypton ion, the mirrors should be first cleaned thoroughly (plus any intercavity component: e.g. etalon, Brew- ster windows or wavelength selectors). For low gain laser systems, such as HeNe or Heed, the clean- ing of the mirrors should only be attempted if they are seriously dirty, since it is very easy to misalign the laser cavity. Be warned--it can sometimes take many hours to realign such a system!!!

Walking in of a Laser

Again using a suitable output power detector (as detailed above) optimize the rear mirror using the vertical and horizontal adjustments. Next select say the top control (vertical adjustment) at the output mirror and turn it slightly clockwise to reduce the output power by about 20-30%. Turn the corresp- ing adjustment on the rear mirror (top control) to maximize the output power--note the reading. Return to the output mirror and again turn the top vertical) control clockwise to reduce the power by 20-30%. Using the rear top control maximize the output power. If the output power increases, then continue this procedure until a maximum is observed. If it reduces, then turn the output mirror control (top) counterclockwise, progressively;, as indicated above, and repeat until a maximum is observed.

Repeat the above for the horizontal controls and finally return and repeat for the top (vertical) adjustments (see diagram below).

CAUTION: Do not rush the above procedure, and beware of any high voltages (lethal) present within the laser head.

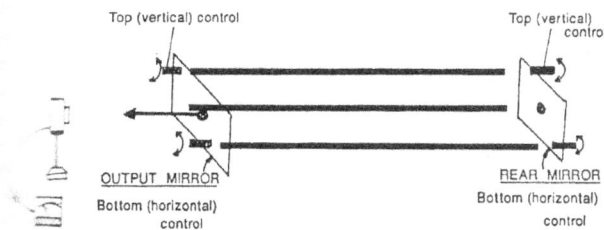

"WALKING-IN" OF LASER CAVITY

Cleaning of Small Mirrors

(Laser cavity mirrors, dielectric and metal coated beam steering mirrors)

The procedure of cleaning small mirrors relies upon the removal of contaminants from the surface by dilution using suitable solvents.

Method: the mirror (or any small component) is held firmly by a gloved hand above a suitable table surface to lessen the possibility of damage if dropped. Any large particles can be blown away using a photographic air- duster (non-CFC type). A piece of soft lens tissue (I have always found the Kodak brand to be one of the best) is placed upon the surface to be cleaned.

The tissue should be of a suitable size to cover the width of the component and be approximately three times longer than the length of the component. High- quality methanol (methyl alcohol; better than AR grade) should be applied with a glass dropper to wet the tissue. Next draw the tissue (direction A) across the surface to remove the contaminants. Repeat this two or three times.

If there are any stubborn residues, then using a piece of lens tissue that is repeatedly folded (2" x 2.5" to size I" x 112") and held using plastic hemostat clamp tongs (the metal type can be used with care--avoid con- tact with component surface). Please ensure that no contaminants from fingers or gloves come into contact with the edge of the folded tissue. Wet the folded tissue with high-grade methanol and flick to remove the excess. Apply the tissue to the surface to be cleaned and gently drag across the surface. Discard the tissue and repeat this until the contaminant is removed. It can be an advantage to repeat the above technique using high- grade acetone. Methanol is the best solvent for removing finger and oil residues.

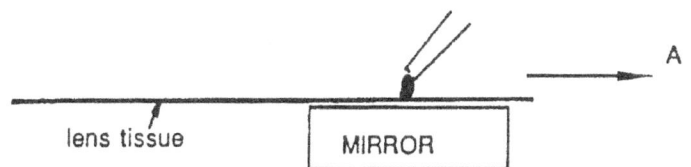

CLEANING OF SMALL OPTICAL COMPONENTS

Other Cleaning Techniques

Large Collimating Mirrors and Lenses

A different technique can be used to clean large reflective surfaces of mirrors and lenses. Normally a collimating mirror is vacuum-coated with aluminium and often with some form of transparent protective coating. These coatings harden with time and offer increased resistance to mechanical contact.

Place the mirror with the reflective surface facing upwards in a container of warm water which has a little mild detergent added. The immersed surface can be very gently swabbed with cotton wool or rubbed care- fully with the fingertips. When all the surface has been treated, transfer the mirror to fresh warm water (no detergent added) and thoroughly agitate the water-- this can be repeated two or three times. Finally, remove the mirror and tilt it on its edge and pour dis- tilled water over the front surface. Then leave to air dry. This method should improve mirror or optics that have become scattery due to dust and finger marks on the surfaces.

CAUTION: Please be very careful, since the coatings are very delicate.

Cleaning Microscopic Objectives

The front surface of microscopic objectives is the surface that usually requires cleaning. This is observed when a laser beam is passed through the objective and projected onto a screen. The use of a lens tissue plus solvent will quickly clean this front surface. Now, if the contaminant has settled on the rear surface, which is recessed within the body of the microscope objective, then this cleaning procedure is a little more involved.

First, remove (if possible) any rear aperture stop (normally screwed into the body). Use an air-duster (non-CFC type) to remove large particles of dust. Next wrap a piece of lens tissue around the soft tip of a cotton bud. Wet with a little methanol and flick to remove any excess. Then push gently into the rear of the microscope objective to make contact with the lens surface, rotate and then remove and discard. Repeat this two or three times and finally repeat using acetone solvent. Do not be tempted to put the complete (assembled) objective into an ultrasonic bath--the lens will tend to fill with the solvent!

Which Laser Shall I Buy? and

Tuning of Laser Optical Systems

written by Ralph Cullen
UK Optical Supplies
84 Wimborne Road West
Wimborne, Dorset
England
(44) 202 886 831 (phone and fax)

12

Recording Materials

There are four different recording media for holography: photopolymer, dichromated gelatin, silver halide, and photoresist. We provide explanations from their manufacturers about their chemical composition and commercial availability. There are also comments from holographers about their use of these materials.

Photopolymer

There are two major producers of photopolymer materials, DuPont and Polaroid.

DuPont

(Directly quoted from DuPont's literature)

DuPont has made significant progress with its line of holographic photopolymer films and replication equipment. OmniDex©-352, a blue-green reflection material, continues to gain momentum and acceptance within the holographic industry as the film of choice, especially for graphic arts products. OmniDex film is currently available in 500' x 12" rolls, 12".

Over the past year DuPont has continued to improve the OmniDex family of films. Two high performance films will become commercial in 1993: HRF-700 and HRF-600. HRF-700 is a high ▲ n blue green sensitive reflection film and HRF-600 is a blue green sensitive high ▲n film for recording transmission or reflection holograms. Feedback from R&D organizations worldwide concerning HRF-600 and HRF-700 films has been positive.

In addition DuPont has also made a significant improvement in full-color and red reflection films.

The new HRF-705X film boasts a film speed of 30 mJ/ cm2 in the red and 20- mJ/cm 2 in the green. This mate- rial is under development but samples are available for evaluation.

Films are available in film thicknesses of 10 and 20 microns.

Processing of DuPont's photopolymer consists of a few simple steps: holographic exposure followed by UV or white-light exposure, optional lamination with a color-tuning film, and heating to brighten the image. The finished hologram is stable to environmental conditions such as heat and moisture.

High-volume replication and processing equipment is also available from DuPont for use with OmniDex© films. The DuPont OmniDex© Replicator operates much like a printing press, using a unique cylindrical scanning geometry for continuous production of holographic images typically at five square feet per minute.

The DuPont Omnidex© Laminator is available with the replicator for color-tuning of rolls or sheets of images, and for downstream conversion operations such as application of pressure-sensitive adhesives or anti-scratch cover films to rolls or individual sheets of holograms. Equipment is also available for heating of large rolls of holograms. The list price of the replication system with the laminator and optics is $310,000.

For further information on DuPont materials or replication equipment, contact the Du Pont Company, Optical Element Venture, Experimental Station, P.O. Box 80352, Wilmington, DE 19880-0352 USA.1r 302- 695-4893. Fax: 302-695-9631.

Polaroid's "Mirage" Photopolymers

Polaroid supplied the image on the front of this book and information about their Mirage© holograms:

There are two ways you can obtain a Polaroid Mirage hologram.

Stock holograms: Choose from selection of already-created holograms. These holograms are available on panels of pressure-sensitive (self-adhesive) labels.

Custom holograms: Hologram images that are determined by the customer. Choose the image or model you would like to capture in a hologram, and we will create it.

Although Polaroid's film is not for sale to the public, Polaroid is more than willing to work with its customers to fill whatever need they may have and offer complete production services from the start of a project t to its completion. For more information call 1-800-237-5519.

Photopolymers in the Desert:

Two Arizona Holographers talk about DuPont's recording material

Dan Norton of Polymer Image in Phoenix, Arizona (USA) has been working with DuPont's materials for about 1.5 years. He previously worked with silver halide and dichromates (about one year with each material).

Although he is getting better results lately, Mr. Norton says it's not because the photopolymer material has improved tremendously since he's been working with it; he has just found better ways to work with it. "It's a lot different than standard holography. Every piece of the technique is different," he says.

Bob Latterman of Polymer Holographics in Jerome, Arizona has been working with DuPont since the beta-testing stage of the new materials. He has been doing small runs and prototypes until now, but anticipates "getting busy in a major way" with photopolymer this year. It's the most important advance In recording materials that I've seen," he says.

Latterman says "I expect to see a lot of activity this year...DuPont should be a significant player in the field. Photopolymer hasn't caught on because people haven't understood it, and DuPont has yet to really promote it for display in consumer applications. (really for display and art markets) Latterman also said that "some people are scared because they think they need a lot of laser power to get started," when in fact he says you can get away with as little as a 100mw at 514nm in green-sensitive film to "get by." He is currently using and Lexel 95 argon and matching krypton laser in the imaging labs, and a Coherent 18 watt for production.

Photopolymer Mastering

The advantages of mastering with photopolymer, says Norton, are its "crispness and clarity of imaging", along with the instant imaging (no need for chemical development as with dichromates and silver halide). He estimates that it takes about 25% of the time to produce a master with photopolymer, as opposed to the other materials. He notes that by using photopolymer "you are closer to real-time interferometry, where you can observe the image as it's being made." In fact, some universities are experimenting with the material in this regard.

Norton also says that the resolution of masters is incredible compared to other materials. He said it "blows away" the "graininess" of silver halide and the "fogginess" of dichromates. He says that photopoly-

mer holograms shot from other types (silver halide and DCG) of masters retain the negative properties of those materials, whereas creating a hologram directly from a photopolymer master results in bright, clear images, with a full range of imaging, even in fluorescent light.

Latterman continues to work in silver halide, but also prefers mastering in photopolymer, although "it's a little bit slower." He reiterated the predictability in processing and also mentioned DuPont's new price structure, which brought the film down from .20/square inch to .05/square inch (very close to the price of silver halide), as two factors that have made photopolymer his choice.

Another advantage, says Norton, is a photopolymer hologram's resistance to moisture and chemicals, a problem with other recording materials.

One disadvantage of the photopolymer, which Norton found out about in the extreme conditions of the American Southwestern desert floor (Phoenix is lower than Jerome by a few thousand feet), is that temperature variations can shift the hologram out of reconstruction range. "If it got above a certain temperature in my lab (100 0   F) I would lose the image," he says. If the 'baking' process is started at this higher temperature, the playback of the hologram itself is shifted.

Latterman had some problems initially with delamination from the tuning film, but mentioned that DuPont has made the film heavier and more stable to correct that flaw. Additionally, they've improved the cover sheet, and "the film feels different now," says Latterman. He attributes the improvements to DuPont's hiring of a full-time staff person devoted to trouble- shooting in the photopolymer field.

"The production of holograms with photopolymer is an incredibly lower cost than dichromate," says Norton. "Because you don't have to work with glass, there is no capping and sealing involved, or cutting of glass." He estimates that the same jobs that used to take 20-30 (people working in dichromates now can be done at his facility with three people.

Dichromated Gelatin (DeG)

Dave Rayfield of US Holographics (formerly Holographic Products) in Logan, Utah has been one of the major players in the dichromate field for several years. He had this to say about working in dichromates.

Sources for the material? You can get dichromate from any major chemical manufacturer, by the barrel. As for previous statements about its short shelf life, Rayfield claims he's had batches around his facility for over a decade.

The biggest improvement he could name in the recent past for dichromates is the ability to do mass-produced color holograms.

While dichromates used to be considered expensive, Rayfield says the supplier problem (Ilford dropped out) in the silver halide field, and the relative expense of photopolymer has made dichromates more accessible. "DCG prices haven't risen," he says. "We used to think that people wouldn't buy them at the prices they are now, but we've found that people buy them readily."

As for alleged difficulties in working with the DCG, Rayfield calls the notion "a myth. The technique is difficult, yes, but it's learned over time." Jerry Heidt, who also works at US Holographics, says "DCG is one of the easiest materials in terms of processing. There is the fix bath, water, and then drying in isopropyl alcohol. It's really quick and doesn't involve any bleaching like silver halide."

Heidt, who works in the mastering process, balked at the notion of "foggy" masters. "Anyone who says that is processing it wrong," he says. "I don't make any foggy masters."

Rayfield did acknowledge the problems of atmospheric differences and moisture exposure to DCGs. "There's nothing that can really be done about it," he says, although some research has been done. He also admitted to the "spectral smear" which can cause a little blurriness in playback of certain frequencies of DCG holograms.

Mastering time for DCG holograms depends on the size of the project and specifications. It can vary from "60 minutes to 6 weeks," says Rayfield. In our last issue (1991) we said that mastering charges for dichromates were "very expensive." Karoline Cullen of Holocrafts, a major dichromate business in British Columbia, Canada, disagrees. "Depending on the size of the order and the type of master being produced," she says, "the fees can be as low as $200.00 or waived entirely."

The most common size that Rayfield's company works with, and the most popular dichromate products continue to be watches. Heidt mentioned sizes of 30mm x 38mm discs, as well as 4" x 5", 5" x 7" and 8" x 10" frames.

Holographic Products uses 5w argon lasers to produce their masters; Rayfield says that the laser power necessary depends on what the dichromate itself is mixed with. Heidt says "you need blue or green (wave- length), but it's not very sensitive to green light." In addition to the 5w argon, Heidt uses both a 40nw argon laser and a Helium Cadmium laser, depending on the size of the hologram.

One of the advantages of DCG is that there is a lack of scatter in blue light, says Heidt. This makes DCG ideal for making Holographic Optical Elements. "No cleaner master for HOEs can be produced than a DCG master," he says (please see our section on HOEs & HUD for more information--several people agree with Heidt on this subject). He mentioned that the photopolymers he worked with from both Polaroid and DuPont had problems with efficiency in the mastering process.

Silver Halide Emulsion

Photography has utilized silver halide emulsions for decades. As you would expect, some major manufacturers of photographic film have adapted their formulae to provide film and plates for holographers.

Availability: There were three primary manufacturers of silver halide films and plates: Ilford, Agfa-Gevaert, and Kodak. The big news in recent times is that Ilford has decided to drop out of the market and no longer supplies holographic emulsions. Agfa-Gevaert has the lion's share of this market.

Advantages: Silver halide requires less light for exposure than most of the other recording materials, so the primary reason for its popularity is its speed. Fast emulsions reduce stability problems. This allows people to buy less powerful, hence less expensive lasers and still make good holograms. It is also convenient to buy silver halide emulsions in precoated film and plates that have a reasonable shelf life, something which cannot always be said of some of the other recording materials.

Drawbacks: The silver halide emulsion is granular, so it scatters more light than the other recording materials.

Commercial Manufacturing Costs using Silver Halide: Before mass manufacturing a hologram you must make a model and have the master hologram made. This cost is one of the things that is holding the industry back. Model costs range from $1,000 to $4,000, depending on the detail, and masters cost between $2,000 and $10,000. Extremely short runs are generally not done because the cost for model and master makes the unit price too high. Although most of the mass manufacturing of silver halide holograms on film and plates is not fully automated, there are fully automated machines for silver halide

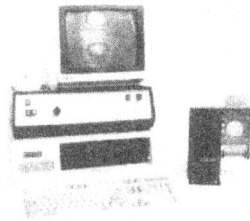

film reproduction which are capable of runs in the tens of thousands. The mass manufacturing quality, which at first was poor, has improved considerably.

Silver Halide Chemistry: There are five atoms which, because of their atomic similarity, are called the halides. They are chlorine, bromine, iodine, fluorine and astatine. Silver halide emulsions are made using either silver chloride, silver bromide, or silver iodide. The other two halides are not used because silver fluoride is insoluble in water and astatine is radioactive.

A typical silver halide emulsion is made by adding a solution of silver nitrate to a solution of potassium bromide and gelatin. Silver bromide crystals form in the emulsion. The emulsion is heated for a certain amount of time, which is called the *ripening process.*

During the ripening process, the grain size increases and the speed of the emulsion is increased. Some doping agents may be added to the emulsion at this time to foster proper crystal growth. Afterwards, the gelatin is allowed to cool. It is then shredded, and the soluble potassium nitrate is washed out of the emulsion.

The emulsion is heated again, with more gelatin added; then it is cooled and applied to a base. The thickness and hardness of the emulsion is important in holography because emulsions too thick tend to deform during development. Emulsions that are too hard can either retard chemical reactions or create vacuoles in the emulsion left by migrating atoms. These vacuoles tend to scatter light.

Exposure Photochemistry: Let's assume our emulsion is made and we now want to expose our emulsion to light. It sounds surprising, but a perfectly structured crystal of silver bromide does not react to light in any appreciable way. A crystal with defects, however, does react with light. Fortunately, most silver bromide crystals will have defects which consist of some interstitial (out of order) silver ions displaced in the crystal structure.

The process of the photochemical reaction is not known in exact detail but it is believed that when light strikes a silver bromide crystal, enough energy is available to remove an electron from an occasional bromide ion. The electron produced is able to migrate through the crystal until it comes in contact with an interstitial silver ion. The silver ion takes the electron and becomes silver metal. Silver atoms formed by this mechanism apparently act as a nucleus for the formation of aggregates of 10 to 500 silver atoms, known as latent images because they are too small to be seen by the naked eye.

After exposure, the emulsion is developed. The developer goes to the site of any silver bromide crystal with a latent image and causes all the silver in that particular silver bromide crystal to be reduced to silver metal and deposited on the already existing latent image of silver metal. This causes a worm-like grain of silver metal to form which is limited in size by the amount of silver available in the silver bromide crystal, This growth is considerable, amplifying the size of the latent image silver metal by a factor on the order of 106, If the developer is left in contact with the emulsion long enough it eventually attacks all the silver in the emulsion. The speed of development is slow enough, though, that you can use a timer to take the emulsion from the developer just after the latent image, but not the unexposed silver bromide crystals, have been developed. At this point the developer has converted silver ions to silver metal if and only if they belong to a silver bromide crystal that was exposed to light. We now place the emulsion in a fixer solution which attacks all silver bromide crystals that were not exposed to light. The fixer makes these silver bromide ions soluble and removes them from the emulsion.

The result is an emulsion with black spots where light has struck, and clear spots where no light struck.

Ideal Silver Halide Emulsion: An ideal silver halide emulsion depends somewhat on its use but there are three main factors to consider in any emulsion: thick- ness of emulsion, grain size of silver halide crystals, an d sensitivity (or density of silver halide crystals) in the emulsion. We can generally state the following:

Thickness of emulsion: It is generally agreed that emulsions of more than 10llm are neither practical or theoretically necessary to produce most volume holo- grams. It is pointed out that thicknesses above this size cause problems in development.

Grain size of crystals: Grain size becomes an important issue in holography because we are recording fringe patterns that are wavelengths apart. Too large a grain size may create excessive scatter which may fog or destroy your hologram and too small a grain size makes the emulsion have no usable sensitivity. It is generally agreed that the most ideal grain size is in the range of.01μm to.035μm.

Sensitivity of emulsion: The ideal exposure would probably be 100 - 300llJ/cm2 to give a useful density (D=2-3). If exposures are much longer than this, the main attraction of silver halide emulsion, its speed, comes into question and other emulsions become more attractive.

Some of the common brands of silver halide emulsions used in holography are listed next, and then we

PLATES	FILM	SPECTRAL SENSITIVITY	EMULSION THICKNESS		GRAIN SIZE
			PLATE	FILM	
Agfa-Gevaert					
10E75	10E75	Red	7μm	5μm	.090 μm
8E56	8E56	Blue-Green	7μm	5μm	.044 μm
8E75HD	8E75HD	Red	7μm	5μm	.044 μm
Eastman Kodak					
649F	649F	All	15 μm	6μm	.060 μm
120-02	SO-173	Red	6μm	6μm	.050 μm
	SO-253	Red	9μm	9μm	.070 μm
	SO-424	Blue-Green	7μm	3μm	.065 μm

(Ilford products for holography have been discontinued)

Note: Soviet and Bulgarian emulsions exist with a grain size of .015 μm but they are not readily available to the market.

Common silver halide emulsions

Some of the common brands of silver halide emulsions used in holography are listed above; next we present recommendations by the manufacturers.

Notes from Manufacturers of Silver Halide Emulsions

1) Agfa-Gevaert

Transmission holography: "While theoretically the maximum diffraction efficiency that may be achieved with an amplitude hologram will be 6.25% at the utmost, theoretically 100% may be achieved with phase holograms.

Technical literature describes a great number of processing systems that, with the highest possible dif- fraction efficiency enable the noise to be kept as low as possible.

The exposure doses that are required for making a good phase hologram amount to -50 Il J/cm2 for the emulsion 8 E 56 HD, and to -25 Il J/cm2 for the emul- sion 8 E 75 HD, as a relatively high density (between D=1.5 and D=2.5) proves to be necessary. As the HOLOTEST 10 E emulsions after bleaching produce more noise than the 8 E types, they are not recommended for phase holography."

Reflection holography: Though theoretically emulsion layers of a thickness of 20 Ilm are necessary for reflection holography so as to achieve reflection holo- grams of top quality, it is still recommended to use the materials 8 E56 HD and 8E75 HD with thickness of the emulsion layer of 7 Ilm,; with these materials the distortion of the Bragg planes after processing will be smaller. This is why it is also possible to achieve high-quality reflection holograms on thinner emulsion layers.

2) Kodak Silver Halide Recommendations: "When selecting a silver halide material, it is important to match peak sensitivity as closely as possible to the wavelength emission of the laser being used. Early attempts to produce holograms also demonstrated the need for emulsions with lowest possible graininess characteristics, and highest possible resolution.

"Another factor to consider is latent image fading, a propensity for microfine-grain emulsions to lose density after exposure; that is, for a given exposure, developed density is likely to decrease with increased time between exposure and processing. Storage conditions between exposure and processing can also contribute to the rate of fading. When conditions do not permit prompt processing, adoption of a uniform protocol for exposure and processing is necessary, especially when each hologram in a series must have the same density."

Photoresist

Photoresist has long been used in the microcomputer industry to stamp out circuit boards. Holographers found out that because of its ability to make an imprint in relief, photoresist was good for creating the shim for stamping out embossed holograms.

There are two options with photoresist: you can buy all the raw materials and coat your own plates, or you can buy pre-made plates. Following is information about the coating process, and then recommendations from manufacturers about using their pre-made plates.

Coating Your Own Plates

If you intend to coat your own photoresist, there is a good article by Burns that you should read because it describes both photoresist and electroplating in detail. There are also worthwhile discussions by Saxby, Bartolini and Clay. The theory of embossed holography has been covered in technical literature as early as 1973.

Emulsion

Most holograms to be embossed are recorded in Shipley AZ1350J emulsion because it is a commercially available positive photoresist emulsion with sufficient resolving power for holography. This photoresist emulsion is designed to be used for the microcomputer industry, however, and consequently does not have the sensitivity that one would wish for making holograms. Hopefully Shipley or a competitor will supply an emulsion more suitable to the holography industry.

Some individuals have tried to make their own resist coating. It is not recommended as resist plates of excellent quality are readily obtainable up to 7 x 7 inches. Emulsions are generally applied by either spin coating or dipping. Spin coating is the preferred method and you can find suppliers that will coat plates for you but you should be very cautious about the results. Your coating should be mirror smooth, free of defects and striations when viewed in yellow safelight.

Exposure:

A frequently used exposure is at 457.9 nm using an argon laser. This is the argon wavelength to which this emulsion is most sensitive but, unfortunately, it is also one of the weakest argon wavelengths. If you use an etalon the available power is reduced by about 50%. At our desired wavelength of 457.9nm, an 18 watt argon with an etalon can put out about 700mw and a 5 watt argon with an etalon produces 120mw. Burns reports the Shipley emulsion is a factor of about ten times more sensitive to the 441.6nm wavelength, which is available from Heed lasers, but Heed lasers are not readily available with more than 40mw of power. In addition, the coherence length of Heed lasers is poor relative to argon lasers equipped with etalons. Therefore, an argon laser with an etalon is preferred for H1s and H2s; whereas Heed lasers are preferred only for contact copies.

To give you an idea of the sensitivity, Burns reports that the Shipley emulsion requires a threshold energy to be effectively exposed. At 457.9nm, it generally requires a total exposure energy of 250,000 ergslcm2 to 2,500,000 ergslcm2 depending on the thickness of the emulsion. Burns found he could achieve this exposure by delivering 150 ergslcm2 per second for a 30-minute exposure or 420 ergslcm2 per second for ten minutes. Either way it is a long exposure and system stability could be a problem.

The exposure Burns finally used delivered 1,000 ergslcm2 per second and the exposure time was 4 minutes 38 seconds. Development was for 40 seconds with tray agitation at a temperature of 53° F using Shipley 303A developer diluted 6:1 with distilled water.

We now coat the photoresist with a layer of electrically conductive silver. Three popular methods for coating and their respective benefits and disadvantages are described below. Regardless of the method, the thickness of the conductive layer only needs to be several hundred angstroms. You want a nice, even thickness with no pinholes or cracks. Watch the temperature to make sure it does not exceed the melting point of the photoresist.

Table 1: Popular Methods for Coating Photoresist w/ Electrically-Conductive Silver

METHOD	COST	SPEED	QUALITY PROBLEMS?
Silver Spray	inexpensive	fast	possible quality problems
Vacuum-deposited silver	expensive	moderate	excellent quality
Electroless nickel	inexpensive	moderate	depends on operator

Silver Spray

This method offers high production rates with low start up costs. It requires only a spray-gun system, spray booth, and solutions. You can use a two nozzle spray gun to mix the two reagents with a commercially available two part silver solution or mix your own solution from existing formulas.

Vacuum-Deposited Silver

To proceed with this method, you need to buy vacuum metallization equipment. In the procedure, the photoresist is affixed at the top of a vacuum bell jar and a small quantity of silver is evaporated from a hot filament in the bottom of the jar. The benefit of this method is that it requires little skill on the part of the operator, and therefore leaves less room for error.

Electroless Nickel

This is a three step immersion process that is inexpensive. It requires dip tanks (one of which must be heated) and solutions that may be mixed in house or purchased commercially. The steps are first to sensitize the photoresist with stannous chloride solution, then dip it in palladium chloride. The third and final step is immersion in electroless nickel deposit, which takes place in a heated tank. There is a wash step between each of the three solutions. The electroless nickel process does take some operator experience since the type of agitation, temperature of solution, immersion time, etc. can all change the outcome.

BUYING PRE-MADE PLATES:

Towne Labs' Manufacturing Process

Towne Labs of Somerville, NJ (USA) is a producer of photoresist plates for use in holography and other fields. They supplied the following description of how their plates are manufactured.

The materials that are used to produce the large Iron-Oxide coated holographic plates are purchased to specification requirements of the micro-electronic, semiconductor and printed circuit board industries for which it was originally designed.

For example, an optical grade polished (both sides) soda lime, float glass substrate 24" x 32" x 0.190" (609.6 x 812.8 x 4.83 mm) has a non-flatness specifica- tion not to exceed 150 x 10 (-6)" per linear inch (1.5 microns per linear cm). Before it is acceptable for Fe02 coating, each piece of glass is cleaned and surface-pol- ished to ensure that the slightest surface imperfections and even micro-dust particles are removed.

The pure Iron-Pentacarbonyl used has a controlled specific gravity of 1.44-1.47 @ 20°C and its deposition is carried out in class 100 clean room conditions. After the Fe02 deposition 16.5 x 10 (-6) inches (0.42 micrometers) thick, the plate is inspected for integrity

of coating. In an area 19" x 27" (482 x 685 mm) no pinholes or imperfections greater than .006 inches (.152) are permitted and only 6 maximum .003 to .006 inches (.076 to .152 mm) are permitted.

The plates are dried in a thermostatically-controlled Class 100 environment, (then) cleaned and inspected again. From that time to the deposition of the photosensitive coating, nothing is permitted to contact the surface of the plate.

The non-carcinogenic Microposit-S-1400-30 highly sensitive photoresist used for coating is spe- cifically formulated to be striation-free. On plates up to 18" x 18", (457 mm square) the photoresist is applied by a spinning process to a final standard thickness of 1.5 +1- 10% micrometers subsequent to a 0.2 micrometer filtration process. This entire pro- cedure is doomed to failure immediately if any por- tion of the image area separates from the glass substrate during exposure, developing, etching, electroplating or when the mother shim is removed from the imaged surface.

The final results are that unless specific attention is devoted to very controlled conditions and to all phases of materials and processes used to produce plates, there is a significant margin for error and the times, effort and expense expended on producing unacceptable material will greatly outweigh the initial investment required to produce quality plates.

COMMENT: The success of the iron-oxide coating is owed primarily to two inherent characteristics of Fe02 coating, e.g. the iron oxide coating effectively absorbs any laser light that may be transmitted through the photosensitive coating. This virtually eliminates light backscatter and the possibility of damage to the primary image. Second, and possibly more important, the iron-oxide coating greatly increases the adhesive quality of the photosensitive coating, thus ensuring the integrity of the imaging and electroplating processes to follow.

Charlie Ondrejik, the marketing manager of Towne, says that of the people who buy their plates, about 20-25% use them for holography, and "the number is growing." A particular area of expansion has been the Far East and China, which Ondrejik called" the most active area at the moment"; they are involved mostly in artistic applications.

Towne has not made any specific modifications to its plates for holographers. "It's a by-product of what we're doing for the semiconductor industry," say s Ondrejik. The semiconductor industry typically uses a plate made with Shipley AZ-1350-J photoresist in small plate sizes up to 9" x 9", whereas holographers use plates of Shipley S-1400-30 photoresist and start at larger sizes of 8" x 10", working up to larger sizes of 24" x 32". The plates used by both industries are coated with the same Iron-Oxide materials, which is vapor-deposited to 3300 angstroms; the difference is in the photoresist material on the plate itself.

Photoresist from Holograpfie-Systeme und Gerate

A new player in the photoresist plate market has arrived recently from Germany: Holografie Systeme und Gerate (HSG). They have provided the following information about their line of products.

"As a result of optimization of the resist composition and the bake- and development-regime we successfully developed a positive photoresist system, which(has) more than twice the sensitivity in the spectral range of 350nm to 500 nm as traditional photoresists. Holographic gratings with diffraction efficiency of up to 90% can be made with this resist. The photoresist REFO-125 is alkaline-developable, filtered to 0.2 um, and striation-free.

The glass plates with the REFO-125 were first developed in HSG especially for applications in diffractive optics.

Processing of photoresist

1. Coating: (resist thickness 1.0 to 2.0 um), e.g. with spin speed of 3000 r.p.m.: 1.5 um

2. Bake 90-95° for 15 min in convection oven.

3. Exposure: e.g. exposure energy at 488 nm:

> 60 mJ/cm2 for high-speed development

$\geq$ 200 mJ/cm2 for high-contrast development

4. Development: immersion in the alkaline developer REFO-125 at 21° C:

e.g. high speed: 30 sec w. the undiluted REFO-D

e.g. high contrast: 60 sec with 5:1 diluted REFO-D, i.e.; 5 parts of REFO-D + 1 part of distilled water.

5. Resist hardening: it is recommended to bake after development in an oven at 95°C for 30 min to increase the resist adhesion for subsequent processes.

6. Resist removal: e.g. with acetone.

Caution: Combustible solvent mixture. Flash point 38°C. Use adequate ventilation. Harmful if swallowed. Avoid contact with skin and eyes. Avoid breathing vapor.

Storage: Store at 10° to 20°C in sealed original con- tainers away from oxidants, sparks and open flame.

Processing of Plates

Exposure: The coated plates can be immediately exposed without a preceding bake.

Development: High-speed variant: immersion in undiluted REFO-D developer for max. 30 sec. high-contrast variant: immersion in diluted REFO-D developer for about 60 sec, for example, 5 parts of REFO-D + 1 part distilled water.

After development the plates should be rinsed with distilled water and blow-dried.

It is recommended to bake the developed plates in an oven at 95°C for 30 min to increase the resist adhe- sion for subsequent processes.

Storage: Coated plates should be kept dry and free of light and dust at temperatures between 10° and 20°. Avoid heat, frost and vapor of organic solutions.

Shelf life: 6 months.

SECTION 5

Names & Addresses

13

Holography Business

This is a list of all businesses and institutions directly involved in holography. When we had to decide who gets listed, we decided to include all individuals and businesses that offer for hire their servieces and products. We know this eliminates some people, like students for example, that are quite knoledgeable about holography, but we had to draw the line somewhere. To broaden our criteria would give us a very large, but not a very useful list.

3D VISION
HOLOGRAMME-LASERPRODUK
Am Dobben 74
Bremen 1
Germany
Voice Phone: (49) 421 76797
Description: Holograms,Holographic
projects. General commercial holograms for
sale and distribution

3DI
49 Upper Woodford
Salisbury
Wilshire
England SP4 6NW
United Kingdom
Voice Phone: (44) 0722-73471
Contact: Caroline Palmer.
Description: Commercial range of holograms.

3M--OPTICS TECHNOLOGY CENTER
1331 Commerce Street
Petaluma
CA 94954
USA
Voice Phone: (707) 765 3240
Contact: Jim Pook
Description: Manufacturer of Holographic
Optical Elements.

A.H. PRISMATIC, INC.
1241 Lerida Way
Pacifica CA 94044
USA Voice Phone: (415) 3552153
Fax Phone: (415) 738 0648
Contact: Deborah Robinson
Description: Distributors of exclusive ranges
of holographic gifts, toys, jewelry, pho- topoly-
mers, and film holograms. Licensed products
available: Star Trek; Star Trek: The Next Generation; Bram Stoker's Dracula; Jurassic Park

A.H. PRISMATIC, LTD.
New England House
New England Street
Brighton
Sussex England BNl 4GH
United Kingdom
Voice Phone: (44) 0273 686 966
Fax Phone: (44) 0273 676 692.
Contact: Ian Dayus
Description: Manufacturers of exclusive
ranges of holographic gifts, toys, jewelry,
photopolymers, and film holograms. Licensed
products available: Star Trek; Star Trek: The
Next Generation; Star Trek: Deep Space Nine;
Jurassic Park

A.N. SEVCHENKO RESEARCH INSTITUT
Minsk
220106
Russia
Contact: V.P. Mikhaylov.
Description: Applied Physical Problems

ABBOTT LABORATORIES
Department 93F
(Building AP-9)
Routes 43 and 137
Abbott Park
IL 60064 USA
Voice Phone: (708) 937 4117.
Contact: Dr. Gerald Cohn.
Description: Holographic Non-Destructive
testing.

ABRAMZIK, CURT
Goethestrasse 67
Offenbach 6050
Germany
Voice Phone: (069) 88 49 11
Description: Conjuring items,Gag items for
festivals and carnival, Holograms. Holo-
graphic projects, Joke items for carnival pare
ties, Magician's items Party supplies,Puzzles
for carnival parties

ACME HOLOGRAPHY
12 Sunset Road
West Somerville
MA 02144
USA
Voice Phone: (617) 623 0578
Contact: Betsy Connors.
Description: Acme Holography is Boston's
first private holography lab. We offer full
service in reflection, transmission and computer generated holography, including design
consultation and large-scale environmental
holography.

AD 2000, Inc.
946 State Street
New Haven
CT 06511
USA
Voice Phone: (203) 624 6405
Fax Phone: (203) 624-1780
Contact: Peter
Description: Fully custom & customized stock
image embossed and photopolymer holo-
grams. Our HoloBank'contains the world's
largest selection of stock image embossed
holograms , available plain, as labels, foil,
magnets, pins, roll stock, etc.

ADEL ROOTSTEIN, INC.
205 West 19th Street,
New York
NY 10011
USA
Voice Phone: (212) 645 2020.
Fax Phone: (212) 929-0342
Description: Sell mannequins used for dis-
plays and holography models.

ADLAS G.m.b.H. & COMPANY KG. Seeland
Strasse 67,
D-2400 Lubeck 14
Germany
Voice Phone: 011+(49)451-390-9300.
Fax Phone: 011+(49)451-390-9399.
Contact: Gerhard Marcinkowski,
Description: Established 1986. Manufacturer
of diode laser-pumped solid state lasers which
operate in CW and pulsed mode with wave-
lengths in IR, visible and UV. Branch office: 4
Craig Road

ADVANCE PHOTONICS
A-147 Ghatkopar Industrial Estate Ghat-
kopar, Bombay 400 086
India
Voice Phone: (91)22582204.
Fax Phone: (91)222024202
Description: Laser distributor.

ADVANCED ENVIRONMENTAL RESCH.
Route 2
Box 2948
Davis
CA 95616
USA
Voice Phone: (916) 757-2567.
Contact: Richard Ian-Frese
Description: Holographic development
applications include architec- tural engineer-
ing, daylighting, PV enhance- ment, energy
conservation, environmental research, and
wavelength selectivity.

ADVANCED HOLOGRAPHIC LABS Lough-
borough University
Ashby Rd.
Louborough
Leicestershire
England LE11 3TU
United Kingdom
Voice Phone: (44) 509 265 232
Contact: Dr. Laurence Holden
Description: see Advanced Holo Labs, USA.

ADVANCED HOLOGRAPHIC LABS (USA)
10440 Ontinveros Place
#1 Santa Fe Springs
CA 90670
USA
Voice Phone: (800) 2554605
Fax Phone: (310) 946 5743
Contact: Francis Tuffy
Description: Worldwide designers, origina-
tors and manufacturers of holographic pat-
tern and image foils and films. See Astor
Universal advertisement for further details.
Offices in Lenexa, KS and Charlotte, NC.

ADVANCED IMAGING TECHNOLOGIES
Normanhurst
7 Broomfield Lane
Hale
Altrincham, Cheshire
England WA15 9 AP
United Kingdom
Voice Phone: (44) 619416367
Contact: Glenn P. Wood, Director.

ADVANCED OPTICS, INC.
5 East Old Shakopee Road
Minneapolis
MN 55420
USA
Voice Phone: (612) 888-0868. Contact: Shari-
lyn Loushin, Description: Manufactureers of
custom and precision optics, first surface mirrors for use in holographic equipment.

AEROSPATIALE.
Ets D'Aquitaine
Saint-Medard-en-Jalles Bordeaux
F-33165 France
Voice Phone: (33)(56) 57 34 8
Fax Phone: (33)-(56) 57 35 48
Contact: C. LeFloc'h
Description: Scientific and   indus-
tri~ research, NDT testing

AEROTECH INC.
Main Headquarters
Electro Optical Division,
101 Zeta Drive,
Pittsburgh
PA 15238
USA
Voice Phone: (412) 963-7470,
Fax Phone: (412) 963-7459,
Contact: Steve A. Botos
Description: Manufacturers of helium nee:.
tubes, power supplies and complete systems
for OEM and end users. Other product line;
include optical table positioners and prec:-
sion rotary and linear positioning systems.
Subsidiary Companies: Aerotech Ltc- Eng-
land, Aerotech GMBH-Germany, Aero- tech
Australia-

AG ELECTRO-OPTICS LTD.
Tarporley Business Centre
Tarporley Cheshire
England CW6 9UY
United Kingdom
Voice Phone: (44) (829) 733305
Fax Phone: (44) 0829 7333679
Contact: Dr. J.A. Gibson
Description: Distributor of lasers, optics, k
equipment.

AGFA
USA Headquarters
100 Challenger Road,
Ridgefield Park
NJ 07660
USA
Voice Phone: (201) 440 2500
Fax Phone: (201) 440 5733.
Contact: Mark Redzikowski
Description: Manufacturers of film, pIa and
recording materials

AGFA LTD
27 Great West Road
Brentford
Middlesex
Egland TW8 9AX
United Kingdom.
Contact: Duncan Croucher
Description: Manufaacturer of film, plates
and recording materials

AGFA N.V.
Corporate Headquarters
Holography Film Dept
Septrestraat 27
Mortsel
B25 10
Belgium
Voice Phone: (03) 444 2111
FAX Phone: (03)444 7094.
Contact: Person: R. De Winne
Description: manufacturers of film, plates and
recording materials

AIMS OPTRONICS SNNV,
Rue Ferd Kinnenstraat 30
B -1950
Kraainem
Belgium
Voice Phone: (32)(027) 310 440.
Fax Phone: (32)(027) 318 918.
Description: Branch office of Newport Corpo-
ration.

AKS HOLOGRAPHIE-GALERIE GmbH.
Potsdamer StraBe 10
Essen 1
4300
Germany
'Voice Phone: (49)(0201) 704 562.
Contact: person: Gudrun Sou,
Description: Embossed hologram manufac-
turer,

ALABAMAA&M UNIVERSITY
P,O. BOX 1268
Normal
AL35762
USA
Voice Phone: (205) 851 5870
Fax Phone: (205) 8515622
Contact: John Caulfield
Description: Scientific holography research,
NDT

ALPHA PHOTO PRODUCTS INC
985 Third Street
P.O. Box 23955
Oakland
CA94623
USA
Voice Phone: (510) 893 1436 ext 1
Fax Phone: (510) 834-8107.
Contact: Maya Fink-Bennett
Description: Wholesale distributor of photo
and graphic arts equipment and supplies

since 1947. International authorized distributor for Agfa holographic film and plates.
Periodic holographic displays in our retail
showroom.

AMAZING IMAGES
10200 West Dodge Rd.
Westroads Mall
Suite 3317
Omaha
NE68114
USA
Voice Phone: (402) 397 2862
Contact: Roger Dorr
Description: Gallery, retail shop, and retail
catalog. We offer the best selection of holog-
raphy images, gifts and novelties in the Mid-
west. Promotional and educational exhibits,
trade shows.Esablished 1990.

AMAZING WORLD OF HOLOGRAMS.
Corrigan's Arcade
Foreshore Road
South Bay
Scarborough
North Yorkshire
England Y011 1PB
United Kingdom
Voice Phone: (44)(0723) 372 857.
Fax Phone: (44) (0482) 492 286
Contact: Carl Racey.
Description: Exhibitors and retailers of film,
glass, embossed, dichromate and related
products. Permanent display of 200 holo-
grams updated and changed regUlarly. Main
season May-October. Distributors of film &
glass.

AMERICAN BANK NOTE HOLOGRAPHICS
Corporate Headquarters
399 Executive Boulevard
Elmsford
New York 10523
USA
Voice Phone: (914) 592-2355.
Fax Phone: (914) 592 3248.
Contact: Edward Dietrich
Description: World leader in develop-
ment of embossed holography for security
and commercial applications. Produces
embossed holograms in a range of formats:
foil, pressure- sensitive & tamper-evident
labels, clear & wide-web laminates for
packaging.

AMERICAN HOLOGRAPHIC INC.
P.O. Box 131
521 Great Road
Littleton
MA01460
USA
Voice Phone: (508) 486 9621
Fax Phone: (508) 486 9080

Contact: Mary Jane Kenney
Description: Design, develop & manufacture
optical components & instruments for use in
industrial & medical measurements. Using
holographic diffraction grating design & mau-
facturing capability to produce components
for unique measurement instruments.

AMERICAN HOLOGRAPHY SOCIETY.
2018 R Street, NW.
Washington D.C
20009
USA
Voice Phone: (202) 667-6322
Fax Phone: (202) 265 8563
Contact: Office of the Secretar
Description: Created to participate in the
development of art. science and technology
of holography and to stimulate and perserve
education, induatry & research.

AMERICAN LASER CORPORATION.
1832 South
3850 West
Salt Lake City
UT84104
USA
Voice Phone: (801) 972-1311.
Fax Phone: (801) 972-5251.
Contact: Dan Hoefer
Description: Established 1970. Manufacturer
rgon, Krypton and Mixed gas laser systems
from 3 mW to 25 W in air or water configura-
tion. Branch Office: HaarGermany .

AMITY PHOTONICS CO.
26 Gibbs Road
Amity Harbor
NY 11701
USA.
Voice Phone: (516) 789-1099.
Fax Phone: (516) 226-8701.
Contact: Paul Westphal
Description: Manufacturers and distributors
of optical lenses, prisms, filters, reticles,
mirror and optical sub-assemblies for all
purposes. Scientific, medical and industrial
. Consulting services available. Established
1985 3 Employees at this address

ANOTHER DIMENSION.
2400 NW Boca Raton Blvd.
#3
Boca Raton
FL33431
USA
Voice Phone: (407) 7501214
Fax Phone: (407) 391 8502
Contact: Stanley Golob
Description: Distributes holograms.

AP HOLOGRAFIKA STUDIO
(Galvanart Bt.)
P.O. Box 113
Budapest
H-1701
Htmgary
Voice Phone: (36-1) 1270412
Fax Phone: (36-1) 1270412
Contact: Tibor Balogh
Description: The AP Holographic Studio
provides mastering, whole process for custom
embossed holograms (mainly in security
applications), optical design using HOEs.
Wholesaler of holographic novelties and dif-
fraction foils.

APPLIED HOLOGRAPHICS, PLC.
Headquarters
Braxted Park, Great Braxted
Malden
Essex
England CM8 3XB
United Kingdom
Description: Holography, stereograms, nar-
row and wide web embossing, hot-stamp foil,
pressure sensitive, polyester, OPP, acetate,
large format holograms, holodisc.

ARBEITSKREIS HOLOGRAFIE B.Y.
Herman-Josef Bianchi
Boeckelter WEG 47
Geldern
4170
Germany
Voice Phone: (49)(2831) 3034
Contact: Christian Liegeois
Description: Artistic holography

ARCHEOZOIC INCORPORATED.
777 Gravel Hill Road
Southhampton
PA 18966
USA
Voice Phone: (215) 3224915
Fax Phone: 215 8620528
Contact: Paul E. Yannuzzi
Description: From product development
to custom work. We present the most
startling images of earth's past, pres-
ent and future through a unique fusion
of model-making and the highest quality
holography.

ARCHITECTURAL GLASS & HOLO-
GRAPH
1 South Street
Great Waltham
Chelmsford
Essex
England CM3 lDF
United Kingdom
Voice Phone: (44) 0245 360431
Contact: Jean Bailey

Description: The design & manufacture of
decorative artchitectural glass, including, on
an experimental basis, holography.

ARMIN KLIX HOLOGRAPHIE
A-Z Schuelldruck
Krouprinzenstr. 126
DUsseldorf
4000
Germany
Voice Phone: (49) 211 317775
Fax Phone: (49) 211 316458
Contact: Armin Klix
Description: Anfertigtmg von Displayholo-
grammen im Auf trag von Werbung und In-
dustrie und Einzel--und Grosshandel. Katalog
von Lieterbaren Hologrammen vorhanden.
Einzelstucke und Grosserien

ART & DESIGN GRAPHICS
Moscow Economic Society
22a, Tverskaya str.
Moscow 103050
Russia, RF
Voice Phone: (95) 299 71 81
Fax Phone: (95) 299 1265
Contact: Oleg B. Serov
Description: Embossed holograms, rare art
holograms, silver halide holograms, reflection
and transmission holograms, master holo-
grams on photoresist, full colour holograms.

ART FREUND HOLOGRAPHY
124 Brookwood Drive
Santa Cruz
CA 95065
USA
Voice Phone: (408) 426 8382.
Contact: Art Freund
Description: Artistic holographer.
ART INSTITUTE OF CHICAGO
Holography Department
Columbus Drive @ Jackson Blvd.
Chicago
IL 60603
USA
Voice Phone: (312) 443-3883
Contact: Ed Wesley
Description: The SAIC offers an MFA
degree with concentration in holography, and
is equiped with three tables, with one con-
taining
a stereogram printer for computer
generated imagry or life. 1 instructor, 3 grad
assistants.

ART, SCIENCE & TECHNOLOGYINST.
Holography Collection
2018 R Street, N.W.
Washington
D.C. 20009
USA
Voice Phone: (202) 667 6322.
Fax Phone: (202) 265 8563.
Contact: L. Bussaut or Odile Meulien.
Description: Established 1983. Research &
educational organization for advancement of
the art, science and technology of holography.
Research focus: 3-D imagery in art and
portrait. Permanent Artwork: Holography
collection. Training sessions.

ARTBRIDGE MANAGEMENT
Mergesstr. 16
Bratmschweig
. 3300
Germany
Voice Phone: (49) 531352816
Fax Phone: (49) 531 352816
Contact: Odile Meulien
Description: :Agency for design & production
of holographic products, services, & commer-
cial art. Custom-made holograms applications
in architecture; in-house innovative products.
Exhibit & educational services including
lectures and training Wholesale/distribution.
Private collection.

ARTIGLIOGRAPHY CO.
7130 Mohawk West Drive
Indianapolis
IN 46236
USA
Voice Phone: (317) 823 0069
Contact: Kerry 1. Brown.
Description: Distribute/display for sale ar
t · work by : Ed Wesly, James MacShane,
Dick Deusterberg, T. Allen Black, Kerry 1.
Brown. Mike Crawford. Custom designer of
holographic displays for all uses. Established
1987. Gallery in Indianapolis.

ARTKITEK
122 Myrtle Avenue
Cotati
CA94931
USA
Voice Phone: (707) 664 2330.
Contact: Steve Anderson
Description: Established in 1986, Artkitek
provides holographic services to the design
community: holograms of architectural mod ·
els, laser sculpture for lobby displays and
special events, HI mastering and purchasing
consultation.

ASAHI GLASS CO.
R & D General Division,
2-1-2 Marunouchi
Chiyoda-ku
Tokyo
100
Japan
Voice Phone: 03 3218 5825
Fax Phone: 03 32145060
Contact: Mr. Fumihiko Koizumi
Description: R&D on holography and single copy holograms.

ASCOT LASER PICTURE STUDIO
27 Upper Village Road
Sunninghill
Ascot
Berkshire
England SL5 7AJ
United Kingdom
Voice Phone: (202) 667 6322.
Fax Phone: (202) 861 0621.
Contact: Mr. Brode!.
Description: Artistic holography; holography education; workshops

ASTOR UNIVERSAL, EAST
11401 Wilmar Blvd.
Charlotte
NC 28241
USA
Voice Phone: (800) 255- 4605
Contact: Don Tomkins
Description: See Advanced Holo Labs, USA

ASTOR UNIVERSAL, MID-WEST
14400 West 96th Terrace
Lenexa
KS 66215
USA
Voice Phone: (800) 255-4605
Contact: Kristina Mikesic
Description: See Advanced Holo. Labs, USA

ATELIER HOLOGRAPHIQUE DE PARIS
13, Passage Courtois
Paris
F-75011
France
Voice Phone: (33)(1) 437969 18
Contact: Pascal Gauchet
Description: Artistic holography; Buying & Selling; Consulting

ATLANTA GALLERY OF HOLOGRAPHY
2960 Coles Way
Atlanta
GA30360
USA
Voice Phone: (404) 352-3412.
Contact: Tanya Scarboro
Description: Only fine art holography gallery in Southeast US. Specializes in showing internationally recognized holographic artists. One-man, group shows exhibited on every-other-month system; permanent gallery also.

AUSTRALIAN HOLOGRAPHICS PTY LTD.
P.O. BOX 160
Kangarilla
South Australia 5157
Australia
Voice Phone: (61) (08) 3837255
Fax Phone: (61) (08) 383 7244
Contact: David Ratcliffe
Description: Manufacturers of white light transmission and reflection holograms, laser transmission holograms, and holographic illumination sources. Specialise in very large format continuous wave holography. Consultancy,
model construction, research and courses available.

ApA OPTICS, INC.
2950 Northeast 84th Lane,
Blaine
MN55449
USA.
Voice Phone: (612) 784 4995.
Fax Phone: (612) 784 2038.
Contact: Anil Jain.
Description: Design and manufacture of head-up displays, HOEs.

BARR & STROUD, LTD.
1 Linthouse Rd.
Glasgow
Scotland G51 4BZ
United Kingdom
Voice Phone: (44)(41) 440 4000
Fax Phone: (44) 414404001
Contact: George Brown.
Description: Artistic holography

BBTINSTRUMENTER APS.
Dronning Olgasvej 6
Frederiksberg
DK-2000
Denmark
Voice Phone: (45)(01)198 208
Fax Phone: (45)(01) 198 747.
Description: Branch office of Newport Corporation.

BEDDIS KENLEY (MACHINERY) LTD .
Unit 3
Elland Terrace
Holbeck Leeds
West Yorkshire
England LS 11 9NW
United Kingdom
Voice Phone: (44)(532) 465 979.
Fax Phone: (44) 532425400
Contact: S.D. Smith.
Description: Large sheet hot foil stamping machines for graphic enhancement and hologram
application. Maximum sheet size 29 x 41 (73cm x l04cm)

BEIJING INSTITUTE OF POSTS
POSTS AND TELECOMMUNICATIONS
Department of Applied Physics
Holography Laboratory
Beijing 10080
China
Voice Phone: (86)(1) 668 1255.
Contact: Hsu Da-Hsiung.
Description: College courses in holography.

BEIJING NORMAL UNIVERSITY.
Analysis and Testing Centre
Beijing 100875
China
Contact: Huang Wanyun.
Description: Industrial and Scientific research, Non-Destructive testing

BENYON, MARGARET
Holography Studio
40 Springdale Avenue
Broadstone
Dorset
England BH18 9EU
United Kingdom
Voice Phone: (44) 202 698 067
Fax Phone: (44) 202 694161
Contact: Margaret Benyon
Description: Independent Holographic Artist.

BERKHOUT, RUDIE
223 West 21st St.
New York
NY 10111
USA
Voice Phone: (212) 255 7569
Contact: Rudie Berkhout
Description: Holographic Fine Artist who has had work exhibited at Whitney Museum of American Art (New York) among other places.

BIAS
(Bremer Institute Applied Beam
Klagenfurter Str. 2,
Bremen 33 2800
Germany
Voice Phone: (49) 0421-218-01
Fax Phone: 0421-218-5063.
Contact: Prof. Dr. lng, or Werner Juptner
Description: Industrial research; holographic
non-destructive testing.

BOB MADER PHOTOGRAPHY.
500 Crescent Court
#160
Dallas
TX 75201
USA
Voice Phone: (214) 8715511
Contact: Hope Hickman.
Description: Artistic holography; pulsed
laser portraits; marketing consultant.

BOBST GROUP
146-T Harrison Avenue
Roseland
NJ 07068
USA
Voice Phone: (201) 2268000.
Contact: Bill Seymour
Description: Manufacturer of hologram
applicator machinery.

BOOTH, ROBERTA
Holographic Artist
5326 Sunset Blvd.
Los Angeles
CA90027
USA
Voice Phone: 213 466 5767
Fax Phone: 213 465 5716
Contact: Roberta Booth
Description: Holographic artist.

BOYD, PJXfRICK
18 Whiteley Rd.
London
England SE19 llT
United Kingdom
Voice Phone: (44) 81 6704160
Fax Phone: (44) 81 6707810
Contact: Patrick Boyd
Description: Holographic Fine Artist

BRAINET CORPORATION
4F Asset Bldg.
3-31-5
Honkomagome
Gunkyo-Ku
Tokyo 113
Japan
Voice Phone: (03) 5395 7030
Fax Phone: (03) 53957029

Contact: Yutaka Inoue
Description: Distrubutor and producer of all
types of film and glass holograms and osther
holographic gifts and stationery goods.

BRANDTJEN & KLUGE, INC,
(Main Office)
539 Blanding Woods Road
Box 736
St. Croix Falls
WI 54024
USA
Voice Phone: (715) 4833265.
Fax Phone: (715) 483-1640.
Contact: Hank A. Brandtjen III
Description: Manufacturer of Hot Stamp
Machinery. Established in 1919. 80 Employ-
ees
at this address.

BRIDGESTONE GRAPHIC
TECHNOWGIES, INC.
375 Howard Avenue
Bridgeport
cr06605
USA
Voice Phone: (203) 366-1595
Fax Phone: (203) 366-1667.
Contact: Rich Zucker
Description: Fully integrated in-house manu-
facturer
of holographic materials and products.
Design, mastering, electroforming,
embossing and all converting. Technology
placement programs.

BRIGHTON IMAGECRAFT.
7 Bath Street
Brighton
East Sussex
England BNl 3TB
United Kingdom
Voice Phone: (44)(0273) 202 069.
Contact: Jeff Blyth.
Description: Specialist in producing holo-
graphic
recording material: red-sensitive
DCG and photopolymer; photochromic plate
for HeNe lasers.

BRITISH AEROSPACE PLC.
Sowerby Research Centre
FPC: 267
P.O. Box 5
Filton,Bristol
England BS12 7QW
United Kingdom
Voice Phone: (44)(0272) 366 842.
Fax Phone: (44)(0272) 363733.
Contact: Mr. S.C.J. Parker.
Description: R&D in HOE and NDT hologra-
phy.

BROOKHAVEN NATIONAL LABORATORY
Upton
NY 11973
USA
Voice Phone: (516) 282 3758.
Description: Industrial research NDT.

BURLEIGH INSTRUMENTS, INC.
Burleigh Park,
Fishers
NY 14453
USA
Voice Phone: (716) 924-9355.
Fax Phone: (716) 924-9072.
Contact: Patty Payne
Description: Burleigh Instruments, Inc. is a
leading manufacturer of Wavemeters and
Fabry Perot Interferometers and Etalon Sys·
terns for laser diagnostics; Piezoelectric·
based micropositioning equipment; Col o~
Center Lasers.

BURNS HOLOGRAPHICS LTD
P.O. Box 377
Locust Valley
NY 11560
USA
Voice Phone: (516) 674 3130
Fax Phone: (516) 6743130.
Contact: Joseph Burns.
Description: Since 1972, Holograms/Stereo-
grams/
Editions; Silver halide, Photores is~
Nickel, Embossed with Agam, Dali, Cos·
sette, Nunez, Dieter Jung, Sam Moree, Ou-.
ers;
1979 - Injection-Molded Holograms
1987 - Design/Development NY % Hologram
CreditCard.

BYRNE, KENNETH
777 8th Avenue
Brooklyn
NY 11215
USA
Voice Phone: (718) 789 5854
Contact: Kenneth Byrne
Description: Fine art holograms.

C.ITOH & COMPANY
Central P.O. Box 136
Tokyo 100-91
Japan
Fax Phone: (81)(3) 639 2946
Description: Holography lab equipment.

CAMBRIDGE LASERS, INC.
853 Brown Rd
Fremont
CA 94539
USA
Voice Phone: (510) 651-0110.
Fax Phone: (510) 651-1690.
Contact: Brian Bohan.
Description: Tube processing service and repairs from re-gassing to complete rebuilding of broken or damaged tubes. Also supply systems complete.

CAMBRIDGE
GROUP.
PO. Box 159
Kendall Square Station
Cambridge
MA 02142-0002
USA
STEREOGRAPHICS
Contact: Stephen A. Benton
Description: Consultants in holographic optics and imaging, and producers of prototype holographic items. Established 1976.4 employees at this address.

CANON INC. R&D HEADQUARTERS
890 , Kawasaki-shi
Saiwai-ku Kawasaki
Kanagawa
211
Japan
Voice Phone: 044 549 5424
Contact: Mr. Tetsuro Kuwayama
Description: Research. Courses in holography

CARLM. RODIAAND ASSOCIATES
P.O. Box 184
3 Locust S1.
Trumbull
cr06611
USA
Voice Phone: (203) 261 1365
Fax Phone: (203) 268 8071
Contact: Carl M. Rodia
Description: Comprehensive engineering consultation services in precision hologram manufacturing. Plant design and engineering, process engineering, troubleshooting and seminar training of manufacturing personnel.
Chapter Thirteen Main Business Listing 121

CASDIN-SILVER HOLOGRAPHY.
(Mailing address)
99 Pond Avenue
Suite 0403
Brookline
MA02146
USA
Voice Phone: (617) 739-6869.
Fax Phone: (617) 739 6869
Contact: Harriet Casdin-Silver.
Description: I have been creating holographic art and interactive holographic installations since 1968. Our company specializes
in original holograms for advertising, architectural and theatre settings, expositions. We are consultants, exhibition organizers/designers.

CENTER FOR ADVANCED
VISUAL STUDIES
Massachusetts Institute of Technology
40 Massachusetts Avenue
Cambridge
MA02139
USA
Voice Phone: (617) 2534415.
Description: Scientific, academic research.

CENTER FOR APPLIED RESEARCH
Art and Technology
72 Lange Boomgaardstraat
Gent
B9000
Belgium.
Voice Phone: (32) (91) 626384.
Fax Phone: (32) (91) 237326.
Contact: Pierre M. Boone
Description: Organization created for positive interaction between science and art, bringing together people interested in visual communication by holography and related techniques .Parent: University Gent, Workshop
Holography, Gent, Belgium.

CENTRAL GLASS CO., LTD.
Glass Technical Planning Dept.
Kowa-Hitosubashi Bldg.
7-IKanda-Nishikicho 3-chom
Tokyo (Chiyoda-ku)
101
Japan
Voice Phone: 03-3259-7354
Contact: Mr. Chikara Hashimoto
Description: Research--Heads Up Display

CENTRAL MICHIGAN UNIVERSITY
Art Department
Mt Pleasant
MI48859
USA
Voice Phone: (517) 774 4280
Contact: Richard Kline.
Description: Artistic holography; holography education; courses/workshops.

CENTRE D'ART
HOLOGRAPHIQUE & PHOTONIQUE
5294 Avenue De L'esplanade
Montreal
Quebec H2T 225
Canada
Voice Phone: Phone (514) 2701840
Contact: Philippe Boissonnet
Description: Artistic services. Non-profit organization. Exhibitions, catalogues, workshops, education.

CFC/APPLIED HOLOGRAPHICS
500 State S1.
Chicago Heights
IL 60411
USA
Voice Phone: (800) GET HOLO
Fax Phone: (708) 758-5989
Contact: Dave Beeching
Description: Fully integrated manufacturer of holograms and security seals. Complete art origination services, in-house manufacturing on wide web and narrow web. Large format (32" x 45"), computer-generated, 2D/3D, dot-matrix and more.

CFC/APPLIED HOLOGRAPHICS CORP
Origination Studio
1721 Fiske Place
Oxnard
CA93033
USA
Voice Phone: 800 438 4656
Fax Phone: (805) 385-5671
Description: Origination and mastering for CFC/ Applied Holographics projects.

CHERNOVTSY STATE UNIVERSITY
2 Kotsyubinsky S1.
Chemovtsy
274012
Russia
Voice Phone: (700) Chemovtsy 44730
Contact: Oleg V. Angelsky,
Description: High-speed holographic and interference methods for non-destructive measurements of surfaces with surface roughness heights 0.001 - 0.5Jlm. Measurements of particle density, size and velocity distribution in disperse media.

CHERRY OPTICAL HOLOORAPHY.
2047 Blucher Valley Road
Sebastopol
CA95472
USA.
Voice Phone: (707) 823 7171
Fax Phone: . (707) 823 8073.
Contact: Greg Cherry.
Description: Highest quality display holography available. Stock and custom reflection! transmission holograms on glass plates or film up to 42" x 72" in size. Limited edition fine art holograms.

CHIBA UNIVERSITY
Faculty of Engineering
Department of Image Science
1-33 Yayoi-cho
Chiba
260
Japan
Voice Phone: 0472 51 1111
Contact: Dr. Kenichi Koseki
Description: Research. Also Miss.Tomoko Sakai at this address for information about Holographic Chemistry, Holography Art

CHIBA UNIVERSITY
Faculty of Engineering,
1-33 Yayoi-cho
Chiba 260,
Japan
Voice Phone: (81)(0472) 511 111 ext
Contact: Jumpei Tsujiuchi.
Description: Scientific research; holographic

CHROMAGEM INC.
573 South Schenley
Youngstown
OH44509
USA
Voice Phone: (216) 7933515
Fax Phone: (216) 793 3515
Contact: . Thomas J. Cvetkovich
Description: Specializing in photoresist masters for mass-production: one-of-a-kind display pieces; consulting and photoresist lab set-ups. Established in 1981. 4 Employees at this address.

CINEMA & PHOTO RESEARCH INSTITUT NIKFI
Prospect 47
Leningradsky
Moscow
Russia
Fax Phone: (7)(1570) 2923
Contact: I. Nalimov.

CISE SPA TECHNOLOGIE INNOVATIVE.
(Mailing address)
P.O.Box 12081 1-20134
Milano
Italy
Voice Phone: (39)(2) 2167 2634
Fax Phone: (39)(2) 21672620.
Contact: Mrs. M. Luciana Rizzi
Description: Various R&D on HOEs and NDT holography.

CITRoEN INDUSTRIE.
35, rue Grange Dame Rose
Meudon-la-Foret
F-92360
France
Contact: Thierry Manderscheid.
Description: Industrial research; holographic non-destructive testing.

CITY CHEMICAL
132 West 22nd Street
New York
NY 10011
USA.
Voice Phone: (201) 653 6900
Fax Phone: (212) 463-9679
Description: Photochemicals and emulsions

COBURN CORPORATION
1650 Corporate Road West
Lakewood
NJ 08701
USA
Voice Phone: (908) 367-5511
Fax Phone: (908) 367-2908
Contact: John White.
Description: Embossing & shim making; training programs.

COHERENT, INC.
Laser Group
5100 Patrick Henry Drive
Santa Clara
CA95054
USA
Voice Phone: (408) 764 4323
Fax Phone: (408) 988 6838
Contact: Tom Hutches
Description: Coherent provides lasers to the industry. Call for catalogue.

COLUMBIA UNIVERSITY.
Department of Otolaryngology,
630 West 168th Street
New York
NY 10032
USA.
Voice Phone: (212) 305 3993.
Contact: Shyam Khanna.
Description: Medical holography: Interferometry and the inner ear.

CONTINENTAL OPTICAL
15 Power Drive
Hauppauge
NY 11788
USA
Voice Phone: (516) 582 3388.
Description: Optics and custom orders.

CONTROL OPTICS
13111 Brooks Drive
Unit J
Baldwin Park
CA 91706
USA
Voice Phone: (818) 813-1990
Fax Phone: (818) 813-1993 .
Contact: W. Liu, President.
Description: Maker of optics and accessories.

CORION CORP.
73 Jeffrey Ave.
Holliston
MA01746
USA
Voice Phone: (508) 429-5065.
Fax Phone: (508) 429-8983.
Contact: Don McLeod
Description: Corion Corp. manufactures volume and one-of-a-kind, custom and stock. optical components including coatings, filters, optics and optical assemblies for use in the UV-Visible-IR spectrum

CORRY LASER TECHNOLOGY INC
414 W. Main Street
PO Box 18
Corry
PAI6407-1728
USA
Fax Phone: (814) 664-3689

COSSETTE, MARIE ANDREE
1145 Avenue des Laurentides
Apt 2
Quebec City
Quebec GIS 3C2
Canada
Description: Holographic fine artist, one-offs & limited editions. Holography exibitions. Private gallery. Holography education, tutoring, studio rental, consultant

COULTER OPTICAL COMPANY
P.O.. Box K
54140 Pinecrest Road
Idyllwild
CA 92349
USA.
Voice Phone: (714) 659 2991.
Contact: Mary Braginton.
Description: Make telescope mirrors, parabolic mirrors and more. Send for free list of poducts.

CRANFIELD INSTITUTE OF TECHNOLOGY
College of Manufacturing
Cranfield
Bedford
England MK43 OAL
United Kingdom
Contact: J.M. Burch.
Description: Scientific, industrial engineering; holographic non-destructive testing.

CREATIVE HOLOORAPHY INDEX
International Catalogue for Holo
Postfach 200210
Bergisch
Gladbach 2
5060
Germany
Voice Phone: Phone (49)
Fax Phone: (49)
Contact: Andrew Pepper, Editor
Description: The Creative Holography Index is an international catalogue, in colour, published four times per year. Available by subscription. It features artists working with holography as a creative medium and includes critical essays. Cost US$55.00/25 pounds Sterling/65 German DM.

CREATIVE LABEL
2450 Estes Drive
Elk Grove Village
IL 60007
USA
Voice Phone: (708) 956 6960.
Fax Phone: (708) 956 8755
Contact: Jerry Koril.
Description: Bindery application of holograms on Kluge (2 stream) and Bobst (4 stream) machines. Call for more information.

CROSS, LLOYD
P.O. Box 672
Gualala
CA954445
USA
Voice Phone: 707- 884 9446
Contact: Lloyd or Cecil Cross
Description: Research in new holographic techniques; holographic fine art works.

CROWN ROLL LEAF, INC.
91 Illinois Ave.
Paterson
NJ 07503
USA.
Voice Phone: (201) 742-4000.
Fax Phone: (201) 742-0219
Contact: James Waitts
Description: Crown Roll Leaf has been supplying embossing material internationally for 5 years. We have been manufacturing shims and embossed products for 4 years. Please call Jim Waitts with any questions .

CSI.
7 Meadowfield Park South
Stocksfield
Northumberland
England NE43 7QA,
United Kingdom
Voice Phone: (44)(661) 842 741.
Description: Manufacture mirrors; optics.

CVI LASER CORPORATION
(Main Office)
200 Dorado Place SE
P.O. Box 11308,
Albuquerque
NM 87192
USA
Voice Phone: (505) 296 9541
Fax Phone: (505) 298 9908
Contact: Bob Soales
Description: Manufactures holographic quality single and multiple element lenses, mirrors, windows, and beam splitters for all standard holographic laser sources. Free 104-page catalog available.

CZECHOSLOVAK ACADEMY OF SCIENCE
Institute of Physics
Na Slovance 2
18040 Prague
8
Czechoslovakia
Voice Phone: (84) 22 41-9 or (84) 2
Contact: Josef Horvath
Description: Holography research

DAI NIPPON PRINTING CO. LTD
Central Research Institute 12,
I-Chome
Ichigaya-Kagacho
Shinjuku-ku
Tokyo 162
Japan
Voice Phone: (81)(03) 266 2310
Contact: Tokio Kodera.
Description: Artistic holography; embossed holography; printing applications.

DAI NIPPON PRINTING CO., LTD.
Business Form Research Laborator
4-5-1 Nishiki-ho
Warabi Ishi
Saitama
335
Japan
Voice Phone: TEL 048442-7 -88
Fax Phone: 04-33-2883
Contact: Mr.Satoshi Yamazaki
Description: Manufacturers. Also
Mr.Shigehiko
Tahara

DAI NIPPON PRINTING CO., LTD.
Central Research Institute
250-1 Aza-Kahasawa
Wakashiba
Kashiwa-city
Chiba277
Japan
Voice Phone: 04 7134 0512
Fax Phone: 0471-33-2540
Contact: Mr. Takashi Wada
Description: Central reserach center. Embossed holography research.

DAIMLER BENZ AG.
Postfach 600202
D-7000
Stuttgart 60
Germany
Contact: H.G.Leis.
Description: Industrial Research; holographic non-destructive testing. HOE research.

DATASIGHTS LTD.
Alma Road
Ponders End
Enfield
Middlesex
England EN3 7BB
United Kingdom.
Voice Phone: (44)(81) 805 4157
Description: Manufacture mirrors for use in holography.

DAVID SCHMIDT HOLOORAPHY.
23962 Craftsman Road
Calwasas
CA91302
USA
Voice Phone: (818) 222-4583.
Fax Phone: (818) 703 1182.
Contact: David Schmidt
Description: Holography courses offered. David Schmidt Holography is a full service mass production laboratory specializing in stereo grams both cylindrical and image plane formats. We also mass produce reflection and transmission holograms for the trade.

DAVIN OPTICAL LTD.
Reliant House
Oakmere Mews
Potters Bar
Hertfordshire
England EN6 9XX
United Kingdom
Voice Phone: (44)(707) 644445.
Description: Manufacture mirrors; optics.

DfuULEEQUWMENTCOMPANY
PO Box 35612
Richmond
VA 23236
USA
Voice Phone: (804) 674 9740
Fax Phone: (804)6749717
Contact: John Williamson.
Description: Manufactures high-precision
holographic embossing machinery; silvering,
electro forming, hot-stamping, laminating,
die-cutting equipment. Embossing, laminat-
ing and plating work for customers.

DB ELECfRONIC INSTRUMENTS S.R.L.
Via Teano 2
Milano
1-20161
Italy.
Voice Phone: (39)(02) 646 934
Fax Phone: (39)(02) 645 6632.
Description: Newport Co. branch office.
Laser supplier.

DE LA RUE HOLOGRAPHICS LTD.
52 Invincible Road
Farnborough
Hampshire
England GU14 7QU
United Kingdom
Voice Phone: (44) 252 520052
Fax Phone: (44) 252 373 871
Contact: P.M.G. Hudson
Description: Integrated manufacturer of holo-
graphic security components for protection of
brands and value documents. Hot stamp foil,
tamper-evident labels & overlaminates from
custom-designed origination

DEC-ART INC
1190 Lavallee
Prevost
Quebec Canada
Voice Phone: (514) 224 8505
Contact: Denis Picard
Description: We carry all the finest holo-
graphic
products. Now exporting to Germany.
European customers please Maria
Erhard at 53-02222-5566.

DEEM, REBECCA
709 1(2 West Glen Oaks Blvd.
Glendale
CA91202
USA
Voice Phone: Phone (818) 549 0534
Fax Phone: (818) 549-0534
Contact: Rebecca Deem
Description: Holographic artist.

DEEP SPACE HOLOGRAPHICS
1070 Moss Street
#105
Victoria
British Columbia V8Z 2Xl
Canada
Voice Phone: (1)(604) 384-3927.
Contact: Karan Wells
Description: Exotic fine art/commercial sculp-
ture/animation. conceptual/industrial design.
display merchandising, exhibits and special
effects. Since 1980 secured worldwide distri-
bution of our DCG disigns via Holocrafts. Star
Trek holograms design.

DELL OPTICS COMPANY, INC
25 Bergen Blvd.
Fairview
NJ07022
USA.
Voice Phone: (201) 941-1010.
Fax Phone: (201) 941-9524.
Contact: Belle Steinfeld
Description: Custom working of precision
optical components. Established 1950. 15
Employees at this address.

DEUTSCHE GESELSHAFT
FOR HOLOGRAFIE
Lerchenstr. 142 a
Osnabriick,
D-4500
Germany
Voice Phone: (49)(0541) 7102
Fax Phone: (49) (0541) 74297.
Contact: Dr. Peter Zec
Description: The society was founded to pro-
mote awareness of holography, and its mem-
bers are mainly holographers and artists. To
this end, the group intends to organise exhibi-
tions.

DIALECfICA AB
Skanegatan 87
6tr
S-11637
Stockholm
Sweden.
Contact: Ambjorn Naeve.
Description: Artistic holography.

DIAURES S.A. HOLOGRAPHIC DIVISIO
Via 1 Maggio 262/ A
1-41019 Soliera
(Modena)
Italy
Voice Phone: (39)(059) 567 274.
Description: Artistic holography; embossed
holography; equipment & supplies.

DIE DRITTE DIMENSION.
Frankfurter StraBe 132-134
Neu Isenburg
D-6078
Germany
Voice Phone: (49)(06102) 33367.
Fax Phone: (49)(06102) 36709.
Contact: Mrs. Elke Hein
Description: Greatest specialized shop for
holography in Germany. Always over 1,0 dif-
ferent holograms in stock. Very comprehen-
sive fine art section. Branch office: Nordwest-
Zentrum, Tituscorso, D-60OC Frankfurt/M.
50, Germany.

DIFFRACTION COMPANY
P.O. Box 151
Riderwood
MD21152
USA
Voice Phone: (410) 6661144.
Fax Phone: (410) 472 4911.
Contact: Hugh C. Wynd
Description: We offer 58 patterns available
in 16 colors with a variety of adhesives Color
explosion graphics/micro-etching an alterna-
tive to 3D; Custom embossing 0: holograms;
Dazzlers-Stickers.

DIMENSIONS
Taj Pura
Sialkot
Pakistan
Voice Phone: (92) 432 66006
Fax Phone: (92) 432 52428
Contact: Mr. Shahjahan
Description: Importers and distributors of all
types of holograms for different applications.

DIRECf HOLOGRAPHICS.
P.O. Box 295
Strasburg
PA 17579
USA.
Voice Phone: (717) 687 -9422.
Fax Phone: (717) 687-9423.
Contact: Jacque Phillips
Description: The exclusive distributor of film
holograms for Third Dimension Ltd. Ove:130
different images in stock. Alsc embossed
sticker, magnets & keychains Quality line of
holographic earrings.

DOMUSU INTERNATIONAL
3F Yoshiba-building
2-17-65 Akasaka
Slato-leu
Tokyo
107
Voice Phone: 03-3585-2238
Fax Phone: : 03-3584-7739
Contact: Mr. Mitsuo Kodera

DREAM IMAGES.
Postfach 1602
Vermeerweg 15
Wesseling D-5047
Germany
Voice Phone: (49) 02236 43138
Fax Phone: (49) 02236-82369.
Contact: Klaus Thielker
Description: Artistic holography; gallery in
Castle of Linl/Rhein since 9/89; marketing
consultant.

DUSTON HOLOGRAPHIC SERVICES
115 Shannon Street
Ottawa
Ontario KIZ 6Y6.
Canada
Voice Phone: (613) 722-9004.
Contact: Deborah A. Duston
Description: Duston Holographic Services
consults corporate and government clients
on HOEs, remotely sensed Holographic Stereograms and the educational and curatorial
aspects of Holography. Deborah Duston is
also a well known artist-holographer.

DUTCH HOLOGRAPHIC LABORATORY.
Kanaaldijk Noord 61
5642 JA Eindhoven
Netherlands
Voice Phone: (31)(40) 817 250.
Fax Phone: (31)(40) 814 865.
Contact: Walter Spierings, Director.
Description: Manufacturer of holoprinter and
holotrack equipment. Production of holograms
on silver halide, photoresist and photopoly-
mer. Computer-generated holograms and
multiple photo-generated holograms (MPGH).
Also traditional recording techniques.

E.C. SCHULTZ & COMPANY,
333 Crossen
Elk Grove Village
IL 6oo07
USA
Voice Phone: (708) 6401190
Fax Phone: (708) 640 1198.
Contact: Bob Schultz
Description: Our company makes stamping,
embossing debossing and applique dies for the

graphic industries. Quality craftmanship and
94 years experience joining for innovative,
distinctive and exciting effects in todays
demanding market.

E.I. DUPONT DE NEMOURS & CO, INC
Optical Elem Venture
Experimental Station
P.O.Box 80352,
Wilmington
DE 19880-0352
USA.
Voice Phone: (302) 695 4893.
Fax Phone: (302) 695 9631.
Contact: Paula Bobeck,
Description: HOE and photopolymer
research at this address.

ElF PRODUCTIONS
EIZYKMAN / FIHMAN.
19 Rue Jean Jacques Rousseau
Paris
F- 75001
France.
Voice Phone: (33)(1)4236063
Contact: Claudine Eizykman.
Description: Artistic holography; mUltiplex.

EALING ELECTRO-OPTICS (UK)
15 Greycaine Road
Watford
Hertfordshire
England
WD24PW
United Kingdom
Description: Supplies optics for holography
labs. Branch office for Ealing Co. USA.

EALING ELECTRO-OPTICS INC.
89 Doug Brown Way
New Englander Industrial Park
Holliston
MA01746
USA
Voice Phone: (508) 429 8370.
Fax Phone: (508) 429 7893.
Contact: Sales Dept.
Description: Manufacturer of mirrors &
optics.

EALING SCIENTIFIC LTD
6010 Vanden Abeele
St. Laurent
Quebec H9R 4N9
Canada
Voice Phone: (514) 3350792
Description: Manufactures lasers.
Chapter Thirteen Main Business Listing 125

EASTMAN KODAK COMPANY.
Scientific Imaging Department
Dept. 841-S
343 State Street
Rochester
NY 14650-0811
USA
Voice Phone: (1) (800) 242-2424
Fax Phone: (1) (716) 7815986
Contact: Donna (HSD) ext 12
Description: Manufacturer of holographic
plates & film.

ED WESLY HOLOGRAPHY.
5331 N. Kenmore Ave
Chicago
IL60648
USA
Voice Phone: (312) 784-1669
Contact: Ed Wesly
Description: Holographic Fine Artist. I am an
artist making candy for the eyes using Ho-
lographic Optical Elements and junk (found
objects).

EDMUND SCIENTIFIC COMPANY.
101 East Gloucester Pike
Barrington
NJ 08007
USA.
Voice Phone: (609) 547 3488.
Fax Phone: (609) 573 6295
Contact: Sales Department
Description: Mailorder catalogue, whole-
sale, and retail We offer one of the largest
selections of precision optics and optical
components and accessories for the optical
lab. Holography products for schools, science
fairs, etc.

ELECTECH DISTRIBUTION SYSMS
605-A, MacPherson Road
#03-03
Citimac Industrial Complex
Singapore
1336
Singapore
Voice Phone: (65)2869933.
Fax Phone: (65) 284 3256.
Contact: Ching-Wat Chia
Description: Supplier of Coherent lasers and
Newport holographic equipment.

ELECTRIC LIGHTING AGENCY
6 West 20th Street
New York
NY 10011
USA
Voice Phone: (212) 645 4580.
Description: Lighting fixtures used by holog-
raphy retailers.

ELECTRO OPTIC CONSULTING SERVICE
18198 Aztec Court
Fountain Valley
CA92708
USA
Voice Phone: (714) 964-0324.
Fax Phone: (714) 536-7729.
Contact: Dr. Colleen Fitzpatrick
Description: PhD Physics: HNDT, laster/optical system design; high-density infonnation storage. Silver halide, thermoplastic holography; laser broker. CW, pulsed laser work. Lab facilities for lease. Very successful proposal writing/submission.

ELECTRO OPTICAL INDUSTRIES, INC.
859 Ward Drive
Santa Barbara
CA93111
USA
Voice Phone: (805) 964 6701
Fax Phone: (805) 967 8590
Contact: Joseph Lansing
Description: Manufacturer of infrared test and calibration instrumentation including: collimators, choppers, blackbody sources, differential temperature sources, FUR test equipment, radiometers and LLL-TV target simulators.

ELECTRO OPTICS DEVELOPMENTS LTD.
Howards Chase
Pipps Hill Industrial Estate
Basildon
Essex
England SS14 3BE
United Kingdom.
Voice Phone: (44) 268 531 344.
Fax Phone: (44) 268 531342
Contact: Mr. Chris Plumb
Description: Equipment & supplies; optics

ELEFPLC
4D E-IF Gelders Hall Road
Shepshed
Leicestershire
England LE12 9NH
United Kingdom
Voice Phone: Phone (44) 509 600220
Fax Phone: (44) 509 508 795
Contact: Peter H.L. Woodd
Description: A total secure service from concept design artwork to finished product-specializing in customer service and delivering quality embossed security and non-security work on time.

ELEKTRO-PHYSIK AACHEN GmbH
Jiilicher Strasse 338
Aachen
5100
Germany
Voice Phone: (49) 241 53 1778
Fax Phone: (49) 241 1822100
Description: Laser optics,Lasers for positioning, Positioning lasers

ELUSIVE IMAGE.
603 Munger Street
316
Dallas
TX 75202
USA
Voice Phone: (214) 720 6060
Contact: Fred Wilbur.
Description: Holography gallery.

EMPAQUES Y EVOLTURAS
HOLOGRAFICAS S.A. DE C.Y.
Pino 343, Local 3
Col. Santa Maria la Ribera
06400
Mexico, D.F.
Voice Phone: (525) 541 1791
Fax Phone: (525) 547-4084.
Contact: Dan Liebennan/Marilu C.
Description: Holography can be used on: packaging, literature inserts, wrapping paper, security labels, bar codes, security paper, stickers, point-of-sale displays. Anything you can imagine can be done in holography.

ENVIRONMENTAL RESEARCH
INSTITUTE OF MICHIGAN (ERIM)
Optical and Infrared Science Lab
P.O.Box 134001
Ann Arbor
Ml48113-4001
USA
Voice Phone: (313) 9941220.
Fax Phone: (313) 994-5704.
Contact: Juris Upatnieks.
Description: Industrial & academic research. BOD & environmental research.

EXCITEK INC.
277 Coit Street
Irvington
NJ07111
USA
Voice Phone: (201) 372 1669.
Fax Phone: (201) 372 8551
Contact: Greg Springer
Description: Supplier of re-manufactured argon and krypton ion laser tubes, and used laser systems. Established in 1984. 10 Employees at this address.

EXPANDED OPTICS LIMITED
Noon Lane
Barnet
Hertfordshire
England EN5 5ST
United Kingdom
Voice Phone: (44)(81) 4412283
Fax Phone: (44) 81 449 6143
Contact: Mr. T.R. Hollinsworth
Description: Manufacturer of medical and industrial endoscopes; micro-precisiooptics.

FANTASM A INC
15 Constitution Way
Woburn
MA 01801-1024
USA
Voice Phone: (617) 938 9910
Fax Phone: (617) 938 9928
Contact: Roger Dreyer
Description: The world's largest manufacturer of 3D holographic watches and consumer products. Sole U.S. distributor of Polaroid films for watches. Complete hologram kiosks available.

FISHER SCIENTIFIC.
E.M.D. Division
4901 West Lemoyne Avenue
Chicago
IL 60651
USA.
Voice Phone: (312) 378 7770.
Fax Phone: (312) 378 7174
Contact: Sales and Marketing Dept.
Description: Supply science lab equipment holography kits, lab manuals, lasers and laser related equipment.

FLASHPOINT DESIGNS INC
23715 West Malibu Road
Suite 230
Malibu
CA90265
USA
Voice Phone: Phone (310) 465 5693
Fax Phone: 2134655716
Contact: Roberta Booth
Description: Holographic design company.

FLATIRON STUDIO
15 West 24th Street
7th Floor
New York
New York 10010
USA
Voice Phone: Phone (212) 929 7938
Contact: Frank Bunts
Description: Painted interference pattern creating depth and movement effects painted canvas.

FLEX CON COMPANY, INC.
Flexcon Industrial Park,
Spencer
MA01562
USA
Voice Phone: (508) 885-3973.
Fax Phone: (508) 885-8400.
Contact: Joseph P. Morgan, Jr.
Description: Manufacturer of holographic and prismatic materials for packaging, gift wrap and graphic film markets. Wide-web embossing in excess of 60 inch width. Range of pressure sensitive films for electronics, medical etc.

FOCAL IMAGE LTD.
P.O.Box 1916
London
England W11 3QR,
United Kingdom.
Voice Phone: (44) (071) 229 0107.
Fax Phone: (44) (071) 727 3438.
Contact: Kaveh Bazargan.
Description: Consultancy in holography; display holograms; holographic optical elements; computer graphics; electronic publishing.

FORD RESEARCH STAFF
Scientific Research Labs (SRL)
20000 Rotunda Drive
Room S-1023
Dearborn
MI 48121
USA.
Voice Phone: (313) 3231539.
Contact: Gordon Brown.
Description: Industrial research; Holographic non-destructive testing. Research and development in computer-aided holographic interferometry.

THE FOREIGN DIMENSION
Suite 1901
Manley Commercial Bldg.
367-375 Queen's Road Central
Hong Kong.
Voice Phone: (852)(542-0282.
Fax Phone: (852) 541-6011.
Contact: Frederic Schvartzman,
Description: Specialists in manufacturing all kinds of holographic and illusion products (Watches, keyrings.), If you are a hologram manufacturer, we can make top quality products at unbeatable prices using your holograms!

FORNARI, DAVID
813 Eighth Avenue
Brooklyn
NY 11215
USA
Voice Phone: (718) 9653956
Contact: David Fornari
Description: Artistic holographer; silver halide transmission & reflection holograms.

FOSTEC GmbH FEINMECHANIK
Opt. Systemtechnik
Halberstaedter Str 7
Berlin 31
1000
Germany
Voice Phone: (49) (030) 8 91 5077
Fax Phone: (49) (030) 891506
Description: Laser technology, components, assemblies and systems,Measuring tables

FOUNDATIONIDEECENTRUM.
P.O. Box 222
5600MK
Eindhoven
Netherlands
Description: Gallery.

FREE UNIVERSITY OF BRUSSELS.
Department of Applied Physics
(ALNA)
Faculty of Applied Sciences
Brussels
Belgium
Contact: Stephan Roose.
Description: Academic and Scientific research.

FRESNEL TECHNOLOGIES INC
101 West Morningside Drive
Fort Worth
TX76110
USA.
Voice Phone: (817) 9267474.
Fax Phone: (817) 926 7146.
Contact: Linda H. Claytor.
Description: Manufactures plastic Fresnel lenses & lens arrays from its POLY IR® plastics for use into the infrared; also other optical products for use into the ultraviolet from acrylic & other plastics.

FRINGE RESEARCH HOLOGRAPHICS
1179A King Street West,
Suite 080
Toronto
Ontario M6K 3C5,
Canada.
Voice Phone: (416) 5352323.
Contact: Michael Sowdon,
Description: Artistic holography; silver halide holograms; pulse portraits; gallery; workshops; travelling exhibit.

FTI JOFFE
Politechnicheskaya 26
Academy of Sciences
St. Petersburg
194021
Russia
Contact: G.A. Sobolev.
Description: Fine Art holography.

FUJI ELECfRIC CO. LTD
,Mecatronics Division
1-12-1 Yuraku-cho
Chiyoda-ku
Tokyo
100
Japan.
Voice Phone: (81)(3) 211 7111.
Description: Manufactures C02 lasers and related equipment.

FUn PHOTO OPTICAL CO.
Ltd. No. 324
l-Chome
Uetake-Machi
Omiya
Japan.
Voice Phone: (81)(04) 866 30111
Fax Phone: (81)(04) 86510521.
Contact: Takayuki Saito.
Description: Industrial research, optics.

FUTITSU LABORATORIES LTD.
Electronic Systems Division
10-1 Wakarniya
Morinosato
Atsugi
243-0
Japan
Voice Phone: 046248 3111
Fax Phone: 046248 3233
Contact: Mr. Takehumi Inagaki
Description: Embossed Hologram Manufacturer.

G.M. VACUUM COATING LAB, INC.
882 Production Place
Newport Beach
CA92663
USA
Voice Phone: (714) 642 5446.
Fax Phone: (714) 6427530
Contact: Dan Coursen
Description: Mirrors and beamsplitters for holographic use.

GALLERIE ILLUSORIA
Schwarztorstrasse 70
Bern
CH-3007
Switzerland
Contact: Sandro Del-Priete.
Description: Gallery featuring holograms.

GALVOPTICS LTD.
Harvey Road
Burnt Mills Industrial Estate
Basildon
Essex
England SS13 1ES
United Kingdom.
Voice Phone: (44)(0268) 728 on
Fax Phone: (44)(0268) 590445.
Contact: R. D. Wale.
Description: Optics; mirrors, lenses.

GARDENER PROMOTION MARKETING
4165 Apalogen Road
Philadelphia
PA 19144
USA
Voice Phone: (215) 849 4049.
Fax Phone: (215) 8494049
Contact: John Gardener.
Description: As the exclusive package goods
marketing representative for Bridgestone
Graphics, we can show you how holography
can be used for problem solving or enhancing
opportunities compatible with your objectives.

GENERALFEROE
1420 45th St.
Emeryville
CA94609
USA
Voice Phone: 510 658 9787
Fax Phone: 510 658 9787
Contact: Jim Feroe
Description: Holographic consulatation,
holographic fine artist.

GENERAL HOLOORAPHICS, INC.
P.O. Box 82247
Burnaby
B.C., V5C 5P7
Canada.
Voice Phone: (604) 435 6654.
Fax Phone: (604) 432 7326.
Contact: Paula Simson/Bemd Simson;
Description: Distributor of dichromate &
embossed gift and jewelry items (Holocrafts),
silver halide wall and desk decor, and photo-
polymer (Polaroid) for the Canadian market.
Custom and stock.

GERALD MARKS STUDIO.
29 West 26th Street,
New York
NY 10010 -100
USA.
Voice Phone: (212) 889
Fax Phone: 2128895994
Contact: Gerald Marks
Description: Working since 1973

GLOBAL IMAGES, INC.
509 Madison Avenue,
Suite 1400
New York
NY 10022
USA
Voice Phone: (212) 7598606.
Fax Phone: (604) 734 2842.
Contact: Walter Clarke.
Description: Manufacturer of holographic
embossing machines; equipment for emboss-
ing.

GRAY SCALE STUDIOS LTD.
4500 19th Street
#294
Boulder
CO 80304
USA
Voice Phone: (303) 442 5889.
Fax Phone: (303) 442 5889.
Contact: George Sivy
Description: Specialists in design and creation
of models and sculptures for holographic
imaging. Consultant services
offered, six years experience, samples avail-
able
upon request. Can also be reached at US
Holographics

GRESSER, E., KG
An der Warth 10
8703 Ochsenfurt
8703
Germany
Voice Phone: (49) (09331) 22 77
Description: Laser measurement techniques,
Lasers, medical

HARIHARAN, P.
Division of Applied Physics
CSIRO
P.O. Box 218
Lindfield
NSW 2070
Australia
Contact: P. Hariharan
Description: Research scientist; author of
Optical Holography.

HARRIS, NICK
711 East 13th Street
Houston
TX 77008
USA
Voice Phone: (713) 8612865
Contact: Nick Harris
Description: Artistic holographer; portraits
and integrals; consulting.

HELlOS HOLOORAPHY INC.
502 Shawnee
Leavenworth
KS 66048
USA
Voice Phone: (913) 758 1000
Fax Phone: (913) 7885106
Contact: Gene Davis
Description: 2000 square foot store/gallery
exclusively for holograms. Plans for 1993
include a larger store.

HELLENIC INSTITUTE OF HOLOGRA ·
PHY
#28 Dionyssou st.
Chalandri
GR-152-34
Greece
Voice Phone: (301) 68 11 803
Contact: Ch. Keranis
Description: Established in 1987, the Insl: ·
tute aims at the overall itroduction and
promotion
of holography in Greece in a:::
aspects: scientific, technical, artistic an ~
media applications. Exhibitions and course,
vocational training,

HIGH TECH NETWORK
Skeppsbron 2
Malmo
S-211 20
Sweden.
Voice Phone: (460(040) 350 75
Fax Phone: (46)(040) 237667.
Contact: Christer Agehall.
Description: Art in holography; securi :;
applications.

HM-Holographie
Sutthauser Str 207
Osnabrtick
4500
Germany
Voice Phone: (49) 541 88957
Description:Holograms,Holographic proj

HOECHST CELANESE CORPORATIOJ\
86 Morris Avenue
Summit
NJ 07901
USA
Voice Phone: (908) 522 7733
Contact: Gunilla Gilberg.
Description: Embossed & artistic hologr:.phy.

HOLAGE
1881 Eighth Avenue
San Francisco
CA94122
USA.
Voice Phone: (415) 564 1840.
Contact: Brad D. Cantos
Description: Fine art holograms; sih halide
holograms.

HOLART CONSULTANTS & REPORT
18 Bonview Street
San Francisco
CA 94110
USA
Voice Phone: (415) 282
Fax Phone: (415) 2824013
Contact: Gary Zellerbach
Description: Holart Consultants publishes
the Holart Report, providing a comprehensive
record of holographic art sales worldwide.
Holart Consultants also offers appraising,
curating, custom creation, and consulting in
holographic marketing and display.

HOLICON CORPORATION.
906 University Place
Evanson
IL60201
USA
Voice Phone: (708) 8661860.
Fax Phone: (708) 491 7955.
Contact: Dr. Hans Bjelkhagen.
Description: Holicon Corporation specializes
in silver halide holograms, pulse or CW, in
particular, portraits. Large-format reflection
or transmission holograms are made as well
as mass production of film holograms.

HOL03
7 rue dGeneral Cassagnou
Saint-Louis 68300
France
Voice Phone: (33)(89) 69 82 08.
Fax Phone: (33) 89 67 7406
Contact: Chambard
Description: Industrial applications of holog-
raphy: shock and vibration, non-destructive
testing, microholography flow visualization,
contouring. R&D: study and development of
new tools for industrial applications.

HOLO GMBH HOLOGRAFIELABOR
OSNABRUCK
MindenerStr.205
Osnabruck
Germany
Voice Phone: (49) 541 7102 173
Fax Phone: (49) 541 7102 176
Contact: Vito Orazem
Description: Holograms up to 1 x 1 m; em-
bossed holography. Holo-design floortiles,
door signs, lamps.

HOLO IMPRESSIONS INC
47-1 Wu Chuan Rd
Wu-Ku Industrial Park
Wu-Ku Shiang
Taipei Hsein
Taiwan
Voice Phone: (886-2) 299 7576
Fax Phone: (886-2) 299-7050
Contact: Daniel C. Hsu
Description: Embossed holography.

HOLO-DIMENSIONS INC
1274 St. Catherine St. East
Montreal
Quebec H2Y 2H2
Canada
Voice Phone: (514) 5232337
Contact: Jean Roy
Description: Artistic holography

HOLO-IMAGES
P.O. Box 626
Ben Lomond
CA 95005-0626
USA
Voice Phone: (408) 3368006
Contact: Tom White
Description: Retail sales of high-quality
holographic imagery--prints, posters,
watches, jewelry, large format fine art.

HOLO-IMAGES, INC.
167 Washburn Road
Briarcliff Manor
NY 10510
USA
Voice Phone: (914) 9418811
Contact: Dr. DeBitetto
Description: Artistic holography.

HOLO-LASER.
6, rue de la Mission
Ecole
25480
Miserey France
Voice Phone: (33) 1 45 315 27
Fax Phone: (33) 1 48 331 702
Contact: Dr. Jean Jouis
Description: Embossed holography and equip-
ment; artistic holography; buying and selling;
education.

HOLO-OR LTD
P.O. Box 1051
Kiryat Weizmann
Rehovot
Israel
Voice Phone: (972) 8 469 687
Fax Phone: (972) 8 466 378
Contact: Uri Levy
Description: Manufactures computer-generated
diffractive optical elements by VLSI techniques.
Catalogue elements and custom designs. Sub-
strates include AnSe, GaAs, various glasses.
DOE work station--dedicated workstation for
element design, mask generation.

HOLO-SERVICE.
Neuensteinerstrasse 19
CH-4153 Basel
Switzerland
Voice Phone: (41) 502 287
Contact: Edgar Bar
Description: Artistic holography.

HOLO-SERVICE.FRIES
Eulerstrasse 55
Basel
CH-4051
Switzerland
Voice Phone: (41) 6122647
Contact: Urs Fries
Description: Artistic holography.

HOLO-SOURCE CORPORATION
21800 Melrose Avenue
Southfield
MI 48075
USA
Voice Phone: (313) 355 0412
Fax Phone: (313) 355 0437
Contact: Lee Lacey
Description: Manufactures fine quality
embossed holograms and colorful diffraction
grating patterns for catalogue and magazine
covers, direct-mail marketing projects, and
point-of-purchase displays.

HOLO-SPECTRA
7742-B Gloria Avenue
Van Nuys
CA91406
USA
Voice Phone: (818) 994 9577
Fax Phone: (818) 9944709
Contact: R. Arkin
Description: Artistic holography consulting;

HOLOCOM HOLOGRAPHIE
13, rue Charles V
Faculte des Sciences et des Tech
F- Paris
France.
Description: Research in holography, Fine art.

HOLOCOR LB.F. PRINTING INC
95 des Sulpiciens
L'Epiphanie
Quebec JOK IIO
Canada
Voice Phone: (514) 5886801
Fax Phone: (514) 588 4898
Contact: Jean-Robert Bernier
Description: We focus our knowledge in what
you want to see: Holographic microengrav-
ing. We devote our energy to what you need:
Performance. Holocor - from electroforming
(shims) to final embossed hologram.

HOLOCRAFT INTERNATIONAL
P.O. Box 152
Lake Forest
IL 60045-0152
USA
Voice Phone: (708) 234 7625
Contact: William Crist, Jr.
Description: Artistic holography, marketing.

HOLOCRAFfS EUROPE LIMITED.
Barton Mill House
Barton Mill Road
Canterbury
Kent,
England en IBY
United Kingdom
Voice Phone: (44) 227 463223
Fax Phone: (44) 227 450399
Contact: Chris Luton
Description: Specialists in manufacture of
dichromate reflection holograms

HOLOCRAFfS: CANADIAN
HOLOORAPHIC DEVELOPMENT
Box 1035
Delta
British Columbia V 4M 3T2
Canada
Voice Phone: (604) 946 1926
Fax Phone: (604) 9461648
Contact: Karoline Cullen
Description: Holocrafts manufactures di-
chromate holograms, offering both stock and
custom production.

HOLODESIGN STUDIES
Rebenstrasse 20
Riehen
CH4125
Switzerland.
Description: Marketing consulting.

HOLOFAR LAB (SRL)
Piazza Acilia No. 3
1nt.3
Rome
00199
Italy
Description: Artistic holography

HOLOFLEX COMPANY
1413 East Old Church Rd.
Urbana
IL 61801
USA
Voice Phone: (217) 684 2321
Contact: Donald Barnhardt
Description: Holographic velocimetray par-
ticle image (HPIV) testing.

HOLOORAM INDUSTRIES
42/44 Rue de Trucy
Fontenay
Sous Bois
94120
France
Voice Phone: (33) 143
Fax Phone: (33) 1 43 9400
Contact: Hughes Souparis
Description: Communication holograms.

HOLOORAM LAND
284 E Broadway
Mall of America
Bloomington
MN55425
USA
Voice Phone: (612) 854 9344
Fax Phone: (612) 8547857
Contact: George Robinson
Description: Retail store specializing in every-
thing holographic. Product range includes art-
work, watches & Jewelry, tshirts, small gift
items and optical novelties. Framing provided
and lighting accessories.

HOLOORAM, THE
P.O. Box 9035
Allentown
PA 18105
USA
Voice Phone: (215) 4348236
Contact: Frank DeFreitas
Description: Free newsletter on holography.

HOLOGRAMM WERKSTATT & GALERIE
GALLERIE FUR HOLOGRAMME
Via Principale 30,CH
Castesegna
7649
Switzerland
Voice Phone: (41) 8241718
Fax Phone: (411) 8241268
Contact: Horst Gutekunst
Description: Creative workshop, develop-
ments, looking for new and attractive ways
for hologram making.

HOLOORAMS 3D
4 Macaulay Road
London
England SW4 OQX
United Kingdom
Voice Phone: (44) 071 6227729
Fax Phone: (44) 071 6225308
Contact: Jonathan Ross
Description: Having worked in holography
since 1978, Jonathan Ross is now a freelance
consultant offering artists representation and
a private gallery space, and commercial
clients epxertise in mass-marketing applica-
tions.

HOLOGRAMS FANTASTIC & ILLUSIONS
North Richmond
Victoria
3121
Australia
Voice Phone: (61) 37296337
Fax Phone: (61) 3 729 6020
Contact: Trevor McGaw
Description: Glass, film and foil (opp, PET

& PVC) 2D, 2D!3D, 3D & multi images &
patterns. Services to painters, hot-stampers,
packaging, label, security marketing, sales
promotion and advertising. Specialists in foil
holography.

HOLOGRAMS INTERNATIONAL
8355 on the Mall
Buena Park
CA90620
USA
Voice Phone: (714) 840 8111
Fax Phone: (714) 8408111
Contact: Dave or Jean Krueger
Description: Distributor of all kinds of holo-
grams to retail stores and wholesale accounts.
We are known for our fast delivery, friendly
consulting and factory-direct prices. Call or
write for quote or catalogue.

HOLOGRAPHC MARKETING, INC.
9250 SW. First Street
Plantation
FL33324
USA
Voice Phone: (305) 474 9965
Fax Phone: (305) 474 9965
Contact: Mark Rapke
Description: Consultant to foreign and domes-
tic corporations on applications of embossed
holography; broker for fine art holograms;
exporter of artistic holograms, jewelry and
novelties.

HOLOGRAPHIC APPLICATIONS
21 Woodland Way
Greenbelt
MD20770
USA
Voice Phone: (301) 3454652
Fax Phone: (301) 345 4653
Contact: Susan St. Cyr
Description: Technical and marketing sevices
for manufacturers of holographic products.
Design consultation, product development,
vendor selection, project management, and
general contracting for endusers of hologra-
phy.

HOLOGRAPHIC CONCEPTS, INC.
1711 S:. Clair Avenue
St. Paul
MN55115
USA
Voice Phone: (612) 698 6893
Fax Phone: (612) 6981619
Contact: Stephen Sugarman
Description: Consulting for educational and
industrial needs, consulting for artists, secu-
rity, textiles. Classes in holography; displa)
design; silver halide mastering, film or glass
plate copies; commercial product researcr.
and design.

HOLOGRAPHIC DIMENSIONS, INC.
16115 SW 117th Avenue
UnitA-21
Miami
FL 33177-1615
USA
Voice Phone: (305) 255 4247
Contact: John I. Ruff
Description: Origination and mass replication of holographic imagery.

HOLOGRAPHIC IMAGES INC.
1301 Dade Boulevard
Yliami Beach
FL33139
USA
Voice Phone: (305) 531 5465
Fax Phone: (305) 5313029
Contact: Larry Lieberman
Description: Limited-edition art holograms: working with artist. Trade pop & trade show holograms: product shots. Portraits: full color, portraits from movie film.

HOLOGRAPHIC INDUSTRIES, INC
P.O. Box 1109
Libertyville
IL 60048
USA
Voice Phone: (708) 680 1884
Fax Phone: (708) 6800505
Contact: Robert Pricone
Description: Designer and operator of retail galleries/goft shops in major shopping centers. We produce our own pulse holographic images, and can obtain nearly any holographic product worldside.

HOLOGRAPHIC LABEL CONVERTING
7626 Executive Drive
Eden Prarie
MN55344
USA
Voice Phone: (612) 934 5005
Fax Phone: (612) 934 7769
Contact: Scott Labelle
Description: Full service capabilities, 2D/3D holography, designing, embossing, hotstamping, precision die-cutting, wide variety of foils. Custom holographic labeling, magnetic holograms, packaging and more ... You think of it, and we can put it together.

HOLOGRAPHIC OPTICS INC
358 Saw Mill River Rd.
Millwood
NY 10546
USA
Voice Phone: (914) 7621774
Fax Phone: (914) 762 2557
Contact: Dr. Jose R. Margarinos

Description: Manufacturer of holographic optical elements, particularly holographic filters, holographic mirror and beamsplitters. Design and manufacture of prototypes.

HOLOGRAPHIC RESEARCH PTY LTD.
Lot 9
Industry Drive
South Tweed Heads
NSW2486
Australia.
Voice Phone: (61) 7524 6625
Fax Phone: (61) 75 543988
Contact: David Toyer
Description: Specializing in volume production of holograms in silver halide and photopolymer. Australasian agent for The Lasersmith. Full production facilities from model-making through mastering to mounting and display preparation.

HOLOGRAPHIC SERVICE
10 via Civerchio
Milan
1-20159
Italy.
Description: Consultant, holograms on packaging material.

THE HOLOGRAPHIC STUDIO
2525 York Avenue
Vancouver
British Columbia V6K lE4
Canada
Voice Phone: (604) 7341614
Fax Phone: (604) 734 2842
Contact: Melissa Crenshaw
Description: The studio produces quality limited edition multi-color reflection holograms. In addition, we have vast experience in the production of single color and achromatic reflection transfers from ruby pulse masters.

HOLOGRAPHIC STUDIOS
240 East 26th Street
New York
NY 10010
USA
Voice Phone: (212) 686 9397
Fax Phone: (212) 4818645
Contact: Jason Sapan
Description: New York's only gallery and commercial holographic lab. Custom and stock holograms. Integral portrait cinematography, mastering, and scan copies from small to large format. Single or mass-produced holograms.

HOLOGRAPHICS (UK) LTD.
32 Lexington Street
London
England WIR 3HR
United Kingdom
Voice Phone: (44) 714378992
Fax Phone: (44) 714940386
Contact: Jon Vogel
Description: Holographic & 3-D multimedia, design origination and production specialists (est 1982) providing comprehensive service for the corporate, retail, & leisure sectors. All types of holograms, 3-D effects produced to brief.

HOLOGRAPHICS NORTH INC.
444 South Union Street
Burlington
VT05401
USA
Voice Phone: (802) 658 2275
Fax Phone: (802) 658 5471
Contact: John Perry
Description: Designers/producers of large format holography up to 44 x 72 inches (1.1 m x 1.8 m). Known worldwide for the highest quality commercial and fne art display work. Design, model building, production, installation and consulting services.

HOLOGRAPHIE KONZEPT GmbH
Koerberstr 3
Frankfurt 50
6000
Germany
Voice Phone: (49) 69 53 1071
Description: Advertising holograms, Holographic
projects.

HOLOGRAPHIE LABOR
H.M.Mielke
Georgenstr 61/R
Miinchen40
8000
Germany
Voice Phone: (49) (89) 2 71 29 89
Description: Holograms, Holographic projects.

HOLOGRAPHY CENTER OF AUSTRIA.
KahlenbergstraBe 6
Wiirmla A-3042
Austria
Voice Phone: (43) 022758210
Fax Phone: (43) 22 75 82105.
Contact: Irmfried Wober.
Description: Our Holography Laboratory, founded in 1985, is the first in Austria. We are the biggest hologram producers in town. We organize exhibitions in Austria and Germany and sell embossed holograms.

HOLOORAPHY INSTITUTE
P.O. Box 24-153
San Francisco
CA94124
USA
Voice Phone: (415) 822 7123
Contact: Jeffrey Murray
Description: Limited editions; holographic art;
consulting; training.

HOLOORAPHY ISRAEL
21 Hakomemiut Str.
Herzlia
46683
Israel
Voice Phone: (972) 052 572387
Fax Phone: (972) 052 570560
Contact: Hameiri Shimon
Description: Holography Israel specializes in
exhibitions-lectures and demonstrations to
pupils and students-advertising, commission,
sales and production of art holograms.

HOLOORAPHY NEWS
1 Erica Court
Wych Hill Place
Woking
Surrey
England GU22 OJB
United Kingdom
Voice Phone: (44) 483740689
Fax Phone: (44) 483740689
Contact: Ian M. Lancaster
Description: Holography News is the inter-
national business newsletter of this industry.
Coverage and distribution is worldwide. Now
in its seventh year, it is valued for its objec-
tivity, depth and analysis.

HOLOORAPHY WORLD CENTER
Art, Science & Technology Institute
800 K Street N.W.
Box 28
Washington
D.C. 20009
USA
Voice Phone: (202) 4081833
Contact: Laurent Bussat
Description: Permanent exhibit; every two
years there is also the Washington Interna-
tional Exhibit of Holography, with awards for
best holography artwork.

HOLOLASER GALLERY
P.O. Box 23386
Dubai
U.A.E.
Voice Phone: 458826
Fax Phone: 452507
Contact: Abdul Wahab Baghdadi
Description: Holography Gallery and holo-
graphic items; laser shows.

HOLOMART
Premium Technology Ltd.
9 Brunswick Centre
London
England WC1N lAP
United Kingdom
Voice Phone: (44) 1 3534212
Fax Phone: (44) 1 353 0684
Contact: Tanya
Description: Buying & selling holograms.

HOLOMAT
741 East Gorham Street
Madison
WI 53703
USA
Voice Phone: (608) 255 3580
Contact: Matt Hansen
Description: Artistic holography and holo-
graphic engineering consulting.

HOLOMEDIA
MUSEUM.
P.O. Box 45012
Drottninggatan 100
10430 Stockholm
Sweden
Voice Phone: (46) 8105465
Fax Phone: (46) 8 107638
Contact: Mona Fosberg
Description: Broker for embossed and artistic
holography; buying & selling holograms;
holography education; gallery.

HOLOMEDIA FRANCE
31 rue de l' etang a L' eau
Rosny
Sous Bois
93110
France
Voice Phone: (33) 1 4894 12 13
Fax Phone: (33) 1 48 55 8839
Contact: Mr. Luigi Castagna
Description: Wholesale and distribution of sil-
ver halide, jewlry and fine art holograms. Two
retail shops in Toulouse and Lyon, France.

HOLOMEDIA INC.
3-15-22, Takaban, Meguro-ku
Tokyo 152
Japan
Description: Sale and distribution of holo-
grams.

HOLOMEX LTD.
4 Borrowdale Avenue
Harrow.
England HA3 7PZ
United Kingdom
Voice Phone: (44) 814279685
Contact: Mike Anderson
Description: Holographic camera design.
Holographic viewer design. Supplier of film
processing kits and safelights.

HOLOMORPH VISUALS, INC.
P.O. Box 1405
Stn. Desjardins
Montreal
Quebec H5B 1H3
Canada
Voice Phone: (514) 3951618
Contact: Kenneth Chalk
Description: Commercial manufacturing of
silver halide holograms; consulting.

HOLOPRESS KG
Steinrader Hauptstr 57a
Lubeck
2400
Germany
Voice Phone: (49) 451496757
Fax Phone: (49) 451497303
Contact: J. Matthiesen
Description: Mass-production of embosseho-
lograms, complete in-house (mastering resist,
electro forming, roller -embosser manufac-
turer of embossing machines, electroforming
systems, master-labs, die/kiss-cutting and
hot-stamping machines.

HOLOPRINT ROSOWSKI
Postfach 116
Lindenau 23
Issum
D4174
Germany
Voice Phone: (49) 2835 1684
Contact: Rosowski
Description: Workshops, embossed & artistic
holography, buying & selling wholesale.

HOLOPRODUCTION.
35 rue Abbatucci
Huningue
68330
France
Voice Phone: (33) 89 69 8208
Contact: J. Striebig
Description: Embossing consultants; mass-
manufacturing; artistic presentation
consul:ants; holography education; medical
research; NDT; lab installation; equipment
and supplies.

HOLOPUBUC UNBEHAUN
Hirschstrasse 84
Wuppertal-2
D-5600
GermaTIY
Voice Phone: (49) 202 84118
Contact: Klaus Unbehaun
Description: Consulting, education, newslet-
ters "Holography 3D Software" and "Alf:'
Reflexionen", fine arts (Holofotografik) boc
"Holo Show International", founding men:ber
"ART-Association for Holography and New
Media."

HOLOS ART GALERIE
4 Place Grenus
1201 Geneva
Switzerland
Voice Phone: (41) 22 325 191
Contact: Pascal Barre
Description: Gallery, retail sales.

HOLOTEC
Studio flir Holographische
Aufnahmen Inh. M. Wagensonner, G
Heilwigstr 19a
München 82
8000
Germany
Description: Holography artist.

HOLOTEC BIRENHEIDE
Am Steinaubach
Steinau
W-6497
Germany
Voice Phone: (49) 6663 7668
Contact: Richard Birenhei
Description: Holograms of any kind

HOLOVISION AB.
Ostharnmarsgatan 69
Stockholm S-11528
Sweden
Voice Phone: (46) (8) 663 9908
Fax Phone: (46) (8) 663 9332
Contact: Jonny Gustafsson.
Description: Specializing in silver halide
holography with pulsed lasers. Denisyuk and
transferred-type reflection holograms up to
30 x 40 cm. Rainbow holograms with pulsed
laser up to 2 x 7 m.

HOWARD SMITH PRECISION OPTICS
61 Lancaster Road
New Barnett
Hertfordshire
England EN4 BA5
Unted Kingdom
Voice Phone: (44) 1441 7878
Contact: Howard Smith
Description: Manufacture mirrors, lenses.

HUGHES AIRCRAFT CO.
Hughes Power Products
P.O. Box 92426
Building RIN500
Los Angeles
CA 90009-2426
USA
Voice Phone: (310) 334 7753
Fax Phone: (310) 334 7756
Contact: John E. Gunther
Description: Hughes' Holographic Products
department, a pioneer in holography since
1974, offers complete development and pro-
duction services for DCG and photopolymer
including HOEs, image mastering, and mass
production/replication on photopolymer

HYOGO PREFECTUALMUSEUM
Curator- Comtempory Art
Kobe-3-8-3 Harada-Dori
Nada-ku Kobe 657
Hyogo ken
Japan
Voice Phone: 07-801-1591
Contact: Mr. Hitoshi Yamazaki
Description: 20th century Art, History of Art
and Holography, Art and Optics, curating a
exhibition of holography into Art.

IDM ALMADEN RESEARCH CENTER
K69/803
650 Harry Road
San Jose
CA95120
USA
Voice Phone: (408) 9271937
Fax Phone: (408) 927 3415
Contact: Glenn Sincerbox
Description: Scientific holography research;
holographic storage.

ICI AMERICAS
Concord Pike
Wilmington
DE 19897
USA
Voice Phone: (302) 575 3087
Description: Optics, HOEs, gratings

ILLINOIS INSTITUTE OF TECHNOLOGY
Mechanical & Aerospace Engineeri
Engineering Building #1
Room 252-B
Chicago
IL 60616
USA
Voice Phone: (312) 567 3220
Fax Phone: (312) 567 7230
Contact: Cesar Sciarnmarella
Description: Holographic interferometry;
industrial holographic research; non-destruc-
tive testing.

IMAC INTERNATIONAL, INC.
1301 Greenwood
Wilmette
IL 60091
USA
Voice Phone: (708) 256 6646
Contact: 1. Kauffmann
Description: Holography marketing consul-
tants.

IMAGEN HOLOGRAPHY, INC
303 Aspen ABC
Suite J
Aspen
CO 81612
USA
Voice Phone: (303) 9258044
Fax Phone: (303) 925 9176
Contact: Alan P. Morterud
Description: Specialized holographic products
for mainstream marketing applications,
including Holotex .. Advanced Holographic
Textiles-a soft, supple, fully washable render-
ing of reflection holograms in both 2D & 3D,
to a variety of fabrics.

IMAGING & DESIGN
1101 Ransom Road
Grand Island
NY 14072-1459
USA
Voice Phone: (716) 773 7272
Description: Designer using holography.

IMEDGETECHNOLOGY
2123 Fountain Court
Yorktown Heights
NY 10598
USA
Voice Phone: (914) 9621774
Fax Phone: (914) 962 1774
Contact: Michael Metz
Description: Research, development and
manufacturing of edge-lit holograms; creative
and innovative holography and optics ap-
plications and problem-solving; consulting;
hologram brokering.

IMPERIAL COLLEGE OF SCIENCE
Optics Section
Blackett Laboratory
London
England SW7 2BZ
United Kingdom
Voice Phone: (44) 71 589511
Contact: J. Dainty
Description: Courses in holography; scientific
holography research; particle measurement.

INFOTECHINTERNATIONAL
Holography Division
3607 West Magnolia Blvd.
Suite 2
Burbank
CA 91505
USA
Voice Phone: (818) 845 7997
Fax Phone: (818) 845 0312
Contact: Robert F. Cranford
Description: Holographic research and
development; developing holographic camera.

INFRARED OPTICAL PRODUCTS, INC.
P.O. Box 3033
South Farmingdale
NY 11735-0664
USA
Voice Phone: (516) 694 6035
Fax Phone: (516) 694 6049
Contact: Barry Bassin
Description: Manufacturer of infrared lenses,
windows, reflectors, beam splitters, computer-
designed IR lens systems and non-linear opti-
cal coatings.

ING.-AGENTUR FUR NEUE
TECHNO LOGIE IN OPTIK UND
PRECISION ENGINEERING
Frickingen 2
D-7771
Germany
Description: Holographic non-destructive
testing; industrial research.

INRAD, INC
181 Legrand Avenue
Northvale
NJ 07647
USA
Voice Phone: (201) 7671910
Fax Phone: (201) 767 9644
Contact: Maria Murray
Description: Manufacturer of nonlinear
materials, harmonic generation systems,
electro-optic and acousto-optic devices and
drivers. Also provides optical components,
assemblies and optical coatings for the UV,
visible and IR.

INSTITUTE OF ELECTRONICS BSSR
Academy of Sciences--Minsk
22 Logoiski Trakt
220841 Minsk
90
Russia
Voice Phone: (7) minsk 65 35 14
Contact: Yuri Morgun
Description: Development and manufacturing
of highly-coherent monopulse lasers and
double-pulse lasers with high spectral radi-
ance based on ruby, YAG, neodymium for
applications in holography, interferometry
and holographic systems.

INSTITUTE OF NUCLEAR PHYSICS
Leningradska obI.
Gatchina
188350
Russia
Contact: A.M. Bekker
Description: Scientific research using HNDT

INSTITUTE OF OPTICAL SCIENCE
Central University
Chung-Li 32054
Taiwan
Voice Phone: (886) 3 425 7681
Fax Phone: (886) 3 425 8816
Contact: Tang Yaw Tzong
Description: HOEs, academic research.

INSTITUTE OF PHYSICS
Ukrainian Academy of Sciences
Prospect Nauki 46
252650 Kiev
28
Russia
Voice Phone: (7) 22 2158
Contact: Vladimir Markov
Description: Artistic and reflection hologra-
phy; research in recording materials.

INSTITUTE OF PLASMA PHYSICS
AND LASER MICROFUSION
P.O. Box 49
Wroclaw
00-908
Poland
Contact: Zbigniew Sikorsky
Description: Academic research

INTEGRAF.
P.O.Box 586
Lake Forest
IL60045
USA
Voice Phone: (708) 234 3756
Fax Phone: (708) 615 0385
Contact: T.H. Jeong
Description: Distribute holographic films and
plates. We also carry pre-packed processing
chemicals, and a variety of stock holograms.

INTERFERENS HOLOGRAFI D.A.
Museum , Gallery, Studio
Halvor Hoels Gt. 6
Hamar N-2300
Norway
Voice Phone: (47) 65 25050
Contact: Olav Skipnes
Description: Ongoing exhibition of Norway's
largest collection of holograms. Makes glass
(mainly reflection) holograms of museum
exhibits. Continuous wave laser.

INTERNATIONAL HOLOGRAM
MANUFACTURERS' ASSOCIATION
Runnymede Malthouse
Runnymede Road
Egham
England TW20 9BD
United Kingdom
Voice Phone: (44) (0) 784 430447

Fax Phone: (44) (0) 784 431923
Contact: Ian M. Lancaster
Description: Promotes the interests of it;
members-full and associate-and the hologTz-
phy industry worldwide. Founder member;
are among the leading manufacturers in
eac'continent.
Current president: David Tion:arsh, Applied
Holographics.

ION LASER TECHNOLOGY INC.
3828 South Main
Salt Lake City
UT 84116
USA
Voice Phone: (801) 262 5555
Fax Phone: (801) 537 1590
Contact: Jeff Smith
Description: Manufacturer of air-coole:
argon lasers.

ISAST/LEONARDO
672 South Van Ness
San Francisco
CA94110
USA
Voice Phone: (415) 431 7414
Fax Phone: (415) 431 5737
Contact: Elizabeth Crumley
Description: Publisher of Leonardo journal
Special issues on holography; electronic
newsletter; holography hotline on MCI
WELL.

ISHII, MS. SETSUKO
#404
1-23 26
Kohinata, Bunkyo-ku
Tokyo
102
Japan
Voice Phone: 03 3945-9017
Fax Phone: 03 3945 9068
Contact: Setsuko Ishii
Description: Holographic Fine Artist.

JAEGER GRAPHIC TECHNOLOGY
lG.T. Holofoil S.A.
20 Avenue des Desirs
Brussels
B-1140
Belgium
Voice Phone: (00) 322 7359551
Fax Phone: 733 1035
Contact: M. Jaeger
Description: Specializes in all kinds of hot
stamp holograms

JAMES RIVER PRODUCTS
800 Research Road
Richmond
VA 23225
USA
Voice Phone: (804) 378 1800
Fax Phone: (804) 378 5400
Contact: Mike Florence
Description: Since 1985,a world leader In holographic machine production. Products include : shutter controls; photoresist plate spinners; dip coaters; complete electro forming facilities; narrow and wide-web embossers; complete holographic origination labs.

JAPAN COMMUNICATION ARTS CO.
Yonezawa Bldg.2F
2 ·37 Suehiro-cho,
Kita-ku
Osaka
530
Japan
Voice Phone: 06 -314 1919
Fax Phone: 06-315 1900
Contact: Ms. Mineko Fukuma
Description: Sales of cards with hologram.

JAYCO HOLOGRAPHICS
29-43 Sydney Road
Watford
Herts
England WD1 7PY
United Kingdom
Voice Phone: (44) 923246760
Fax Phone: (44) 923 247 769
Contact: Rohit Mistry
Description: Complete production service for embossed holograms. Embossing masters through to fully finished product. Sixteen years of experience enables Jayco to offer outstanding quality of product and service at completitive prices.

JODONINC.
62 Enterprise Drive
Ann Arbor
MI48103
USA
Voice Phone: (313) 7614044
Fax Phone: (313) 7613322
Contact: John Gillespie
Description: Manufacturer of HeNe lasers, laser systems, specialty laser tubes, optical and electro-optical instruments and systems. Holographic films, plates and chemicals. Engineering services.

JOHN HOPKINS UNNERSITY
Dept of Physics and Astronomy
Baltimore
MD21218
USA
Voice Phone: (410) 5167385
Contact: Homaira Akbari.
Description: Scientific holography research;

JOURNAL OF LASER APPLICATIONS
4143 Merriweather Road
Toledo
OH43623
USA
Voice Phone: (419) 885 4803
Fax Phone: (419) 885 5895
Contact: Jack Dyer
Description: Quarterly peer-review technical journal for the laser industry.

JR HOLOGRAPHICS
Suite 1660
100 Wilshire Blvd.
Santa Monica
CA 90401-1135
USA
Voice Phone: (310) 393 2388
Fax Phone: (310) 3938611
Contact: Judy Roberts
Description: Acts as a licene for celebrity images of both company logos and individuals. We expect to place holographic images in major retail markets and product packaging areas.

K.C. BROWN HOLOGRAPHICS
22 St. Augustine's Road
Camden Town
London
England NW1 9RN
United Kingdom
Voice Phone: (44) 0714822833
Contact: Kevin C. Brown
Description: Pulse portraits; artistic holography.

KAC, EDUARDO
725 W. Melrose
#2F
Chicago
IL60657
USA
Voice Phone: (312) 8714619
Fax Phone: (312) 8714619
Contact: Eduardo Kac
Description: Fine art holograms.

KAISER OPTICAL SYSTEMS, INC.
P.O.Box 983
371 Parkland Plaza
Ann Arbor
MI48106
USA
Voice Phone: (313) 6658199
Contact: James McNaughton
Description: Holographic Optical Elements; HUD's.

KARAS STUDIOS S.L.
Hospital, 12
Madrid
28012
Spain
Voice Phone: (34) 1 530 89

Fax Phone: (34) 1 530 8988
Contact: Ramon Benito
Description: Established in 1988; art exhibitions, art gallery, private collection.

KAROLINSKA INSTITUTET
School of Dentistry
Box 4064
Huddinge
S-14104
Sweden
Voice Phone: (46) 8 774 0080
Contact: Hans Ryden
Description: Holography research applied to dentistry.

KAUFFMAN, JOHN
20 Buena Vista Avenue
Point Reyes Station
CA94956
USA
Voice Phone: (415) 663 1216
Contact: John Kauffman
Description: Holographic Fine Artist.

KEIO UNNERSITY
Department of Electrical Enginee
3-14-1
Hiyoshi Kohoku-ku
Yokohama
223
Japan
Voice Phone: 045-563-1141
Fax Phone: 045 563 3421
Contact: Dr. Masato Nakajima
Description: Research using HNDT

KENDALL HYDE LTD.
Kingsland Industrial Park
Stroudley Road
Basingstoke
Hants.
England ,RG24 OUG
United Kingdom
Voice Phone: (44) 0256 840 830
Fax Phone: (44) 0256840443
Contact: M. Kendall
Description: Optical coating specialists.

KEYSTONE SCIENTIFIC CO.
P.O. Box 22
Thorndale
PA 19372
USA
Voice Phone: (215) 3848092
Fax Phone: (215) 384 8093
Contact: Ed Kelly
Description: Manufacturer of film and plate processors and transports. Manufacturer of holography kits. Distributor for AGFA and Kodak holographic films, plates and chemicals.

KINETIC SYSTEMS
20 Arboretum Road
P.O. Box K
Roslindale
MA 02131
USA
Voice Phone: (617) 522 8700
Fax Phone: (617) 522 6323
Description: Manufacturers of Vibraplane
standard and special Honeycomb optical
tables in three grades up to 5' x 12' x 24"
Larger sizes available by butt splicing. Also
vibration isolation support systems.

KREISCHER OPTICS, LID.
906 N. Draper Rd.
McHenry
IL 60050
USA
Voice Phone: (815) 344 4220
Fax Phone: (815) 344 4221
Contact: Cody Kreischer
Description: Custom Manufacturer of master
and production test glasses, optical flats, lens-
es, condensers, cylinders, windows, filters,
prisms, mirrors, beamsplitters, substrates,
magnesium fluoride coatings. Consulting
services in optical design.

KYOTO INSTITUTE OF TECHNOLOGY
Dept. of Electronics and Informa
Matsugasaki
Sakyo-ku
Kyoto
606
Japan
Voice Phone: 075 791-3211
Fax Phone: 075 723 2853
Contact: Toshihiko Kubota
Description: Color Holography, Hologrphic
Optical Element

KYOTO TECHNICAL UNIVERSITY
Dept. of Photographic Technology
Matsugasaki
Sakyo-ku
Kyoto
606
Japan
Contact: Toshihiro Kubota
Description: Artistic holography; DCG, color,
reflection holograms.

L.A.S.E.R. CO.
1900 Gore Drive
Haymarket
VA 22069
USA
Voice Phone: (703) 754 2526.
Contact: Jim Bowman.
Description: Fine art holograms; lighting
consultant.

L.A.S.E.R. NEWS
Laser Arts Society for Education
P.O. Box 24-153
San Francisco
CA 94124
USA
Voice Phone: (415) 822 7123
Contact: Patty Pink
Description: Volunteer staffed non-profit or-
ganization dedicated to holography and laser
education and research. Members receive the
quarterly L.A.S.E.R. News. Two year mem-
bership USA $40, outside USA $60.

L.I.R. E.R.A.
12, rue Libergier
Reims
F-51110
France
Voice Phone: (33) 26884452
Contact: Michel Grosmann
Description: Scientific research; HOEs

LAB CENTRAL DES PONTS ET
CHAUSSE
DES PONTS ET CHAUSSEES (LCPC)
58 Boulevard Lefebvre
F-75015 Paris
75015
France.
Voice Phone: (33)(1) 4532 3179.
Contact: Jean-Marie Caussignac.
Description: Industrial research; holographic
non-destructive testing.

LABOR DR. STEINBICHLER
ArnBauhof4
Neubeuern
D 8201
Germany
Voice Phone: (49) 80351018.
Fax Phone: (49) 8035 1010.
Contact: Dr. H. Steinbichler,
Description: Holographic investigations,
developments on contract basis; application
laboratory for: vibration analysis, nondestruc-
tive testing, deformation measurements, con-
tour measurements, image analysis; pulsed
and CW-lasers, motor test bench, computer
based evaluation.

LABOR FUR HOLOGRAFIE
Arn Forst 38
Wesel
D-4230
Germany
Voice Phone: (49) 281) 52837.
Contact: A. Fuchtenbusch.
Description: Artistic holography; holography
education; fine art holograms.

LABORATORY SOETE
Holography Workshop
University Gent
41 St. Pietersnieuwstraat
Gent
B9000
Belgium
Voice Phone: (32) 91 643242
Fax Phone: (32) 91 237326
Contact: Pierre Boone
Description: Consultancy, education, problem-
solving for display holography. Museum
applications and (mainly!) non-destructive
testing.

LAKE FOREST COLLEGE
Center for Photonic Studies
Holography Workshops
Sheridan and College Road
Lake Forest
IL60045
USA
Voice Phone: (708) 234 3100
Fax Phone: (708) 234-6487
Contact: Tung H. Jeong.
Description: Each summer during July, Lake
Forest College offers a 5-day hands-on
workshop for participants who have no prior
experience in holography. An advanced 5-day
workshop follows. Write for information.

LAMBDA ANALYTICAL LABORATC
RIES
55 Webster Avenue
New Rochelle
NY 10801
USA.
Voice Phone: (914) 654
Description: Holographic non-destructive
testing

LAMBDA/fEN OPTICS
c/o Optical Corporation of Arne
One Lyberty Way
Westford
MA01886
USA
Voice Phone: (508) 692-8140.
Fax Phone: (508) 692 9416.
Contact: George Olmsted.
Description: Products: Precision, large
aperture (to 36 inch diam.) aspheric mirrors
holographic production systems.

LAMINEX/HIGH TECH UK LTD.
Bromfield Industrial Estates
Mold
Clwyd
England
CH7IJR
United Kingdom

Voice Phone: (44)(0352) 58444
Contact: Keith Green.
Description: Artistic holography; security
graphic applications

LASART LID.
P.O. Box 703,
Norwood
CO 81423
USA
Voice Phone: (303) 327 4701
Fax Phone: (303) 327 4701
Contact: Steven Siegel
Description: Lasart, Ltd. specializes in custom DCG work, from modelmaking, mastering and quality finishing. This includes
production and limited edition jewelry,
Swiss watches, and medium format composite
sculpture.

LASER AFFILIATES
2047 Blucher Valley Road
Sebastopol
CA 95472
USA
Voice Phone: (707) 823
Fax Phone: (707) 823 8073
Contact: N. Gorglione.
Description: Laser Affiliates is an award-winning non-profit organization that designs
innovative holographic and laser theatrical
productions, installations and exhibitions.
Services include curatorial guidance, videotapes and media lectures.

LASER APPLICATIONS, INC.
(DIVISION OF LASERMETRICS INC)
12722 Research Parkway
Orlando
FL 32826
USA
Voice Phone: (407) 380 3200.
Fax Phone: (407) 381 9020.
Contact: person: Joseph Salg
Description: Holographic non-destructive
testing; manufacturer ruby/yag lasers; HOE
manufacturer; Holography equipment.

LASERARTS.
1712 Cathedral Street
Plano
TX 75023
USA
Voice Phone: (214) 4230158.
Contact: M. Talbott.
Description: Holographic consultants and
implementers. Commercial utilization of
holography, trade shows, unique promotions
and museum exhibits (design, build, rent or
sell). Venture capitalists consultants. Professionals in business, art, technology and
applications.

LASER ELECTRONICS PTY., LTD.
P.O. Box 359
Southport
Queensland 4215
Australia.
Voice Phone: 61 75532066
Fax Phone: 61 75533090
Contact: N. Walden
Description: Laser Electronics designs and
manufactures an extensive range of lasers
and laser systems across seven industry categories including scientific, educational, and
research units. Custom systems can also be
developed.

LASER FOCUS WORLD
1 Technology Park Drive
P.O. Box 989
Westford
MA01886
USA
Voice Phone: (508) 692 0700.
Fax Phone: (508) 692 0525.
Description: Laser trade magazine; annual
catalogue.

LASER GRAPHICS.
5, Cotta Street
Thessaloniki
55337
Greece.
Voice Phone: (30)(031) 908 087.
Contact: Yannis Palamas.
Description: Artistic holography.

LASER INSTITUTE OF AMERICA
Education Division
12424 Research Parkway
#125
Orlando
FL32826
USA
Voice Phone: (407) 380 1553
Fax Phone: (407) 380 5588
Contact: Jackie Thomas
Description: Laser safety courses. Publishes
Journal of Laser Applications. Hosts annual
International Congress on Applications of Lasers and Electro-Optics (lCALEO), including
holographic applications.

LASER INTERNATIONAL
19 Normanton Rise
Holbeck Hill
Scarborough
N. Yorks
England YOll 2XE
United Kingdom.
Voice Phone: (44)(0723) 366 096.
Contact: Keith Dutton.
Description: Holography Gallery.

LASER IONICS INC.
(Main Office)
701 South Kirkman Road
Orlando
FL32811
USA
Voice Phone: (407) 298 1561.
Fax Phone: (407) 297 4167.
Contact: Drew Nelson
Description: Manufacturer of gas ion lasers
including Argon, Krypton and mixed gases.
Specializing in high power requirements
needing stable power in a compact package.

LASER LIGHT DESIGNS
2412 Kennedy Way
Antioch
CA94509
USA
Voice Phone: (510) 754 3144.
Contact: Michael Malott
Description: New product designs using embossed foil, tinsel and holographic films. I specialize in jewelnd novelty designs. Designer
of the original Rainbow Flasher, inventor of
original Rainbow Sparkler.

LASER LIGHT EXPRESSIONS PTy. LTD
1/5 Gibbons Street
Telopea
New South Wales 2117
Australia
Voice Phone: (612) 890 1233
Fax Phone: (612) 891243
Contact: Rosemary Sturgess
Description: Since 1984, we have been producing commercial holograms and diffractions
for security and display applications. We offer a complete service from artwork creation
through to application & printing within
Australasia.

LASER LIGHT IMAGE.
229 Filey Rd
Scarboro
England YO 11 3AE
United Kingdom
Voice Phone: (44)(0482) 26744
Fax Phone: (44)(0482) 492 286.
Contact: Carl Racey.
Description: Artistic holograms and equipment.

LASER LIGHT LTD.
57 Grand Street
New York
NY 10013
USA
Voice Phone: (212) 226 7747.
Contact: Abe Rezny.
Description: Artistic holography.

LASER OPTICS, INC.
111 Wooster St.
Bethel
CT06801
USA
Voice Phone: (203) 744-4160.
Fax Phone: (203) 798-7941.
Contact: Henry Louis
Description: A complete line of laser and
optical components for ultraviolet, visible
and infrared applications from 250 nm to 16
microns, including focusing lenses, windows,
cavity components, prisms, beamsplitters,
mirrors and coatings.

LASER PHOTONICS, INC.
12351 Research Parkway
Orlando
FL32826
USA
Voice Phone: (407) 281-4103.
Fax Phone: 407 380 3749
Contact: Steve Qualls.
Description: Manufacturer of carbon dioxide
& ruby lasers; equipment & supplies for
holography.

LASER RESALE INC.
54 Balcom Road
Sudbury
MAO1776
USA.
Voice Phone: (508) 443 8484.
Fax Phone: (508) 443 7620
Contact: Jack Kilpatrick
Description: Laser Resale provides a market-
place for buying and selling pre-owned lasers,
laser systems and associated equipment. Cur-
rently available holographic lasers are He:Ne,
15-70 mW, and, argon, 100 mW -20W.

LASER TECH INDUSTRIES
3173 Texas Avenue
Simi Valley
CA93063
USA
Voice Phone: (818) 583-3406.
Fax Phone: (818) 889-5605
Description: Supplies complete support, parts
and service on all lasers. Also manufactures
mirrors, lenses, rods and other parts for
lasers.

LASER TECHNOLOGY, INC.
1055 West Germantown Pike
Norristown
PA 19403
USA
Voice Phone: (215) 631
Contact: Tom Gleason
Description: Manufacture equipment for
laser-based NDT; Holography and Shearogra-
phy equipment and inspection services.

LASERFILM ECKHARD KNUTH
Multi-Plex Holographie
Milchstrasse 12
Munich
D-8000
Germany
Voice Phone: (49) 89 498 714
Contact: Eckhardt Knuth.
Description: Artistic holography.

LASERION HANDELS GMBH
Postfach 110268
Bremen 11
2800
Germany
Description: Artistic holography; commis-
sions.

LASERMEDIA.
6833 Arizona Circle
Los Angeles
CA90045
USA.
Voice Phone: (310) 8203750.
Fax Phone: (310) 338 9221.
Description: Install laser light show exhibi-
tions.

LASERMET LIMITED
Five Oaks
SwayRoad
Brokenhurst
Hants
England S04 27RX
United Kingdom
Voice Phone: (44) 590 23075.
Contact: Dr William F. Fagan.
Description: Holographic Non-Destructive
test

LASERMETRICS, INC.
196 Coolidge Avenue
Englewood
NJ 07631
USA
Voice Phone: (201) 894-0550.
Contact: Robert Goldstein.
Description: Laser manufacturer & Industrial
research

LASERSMITH, INC.
1000 West Monroe Street
Chicago
IL 60607
USA.
Voice Phone: (312) 733 5462.
Fax Phone: (312) 733 5926
Contact: Steven L. Smith.
Description: Specialists in holographic imag-
ing . In-house service: art work origination,
photo shoots of art/object, 2D/3D full color 2D
& 3D mastering. Stereogram filming; separa-
tions & mastering; computer modeling &
rendering for full color.

LASERWORKS
P.O. Box 2408
Orange
CA92669
USA
Voice Phone: (714) 832 2686.
Contact: Selwin Lissack.
Description: Holographic artist.

LASING S.A.,
Marques de Pico Velasco
64
E-28027
Madrid
Spain
Voice Phone: (34)(01) 2683643
Description: Branch office of Newport Corpo-
ration

LASIRIS INC.
Main Office
3549 Ashby
Ville St. Laurent
Quebec
H4R 2K3 Canada
Voice Phone: (514) 335 1005
Fax Phone: (514) 335 4576
Contact: Alain Beauregard
Description: Embossed holography; artistic
holography

LAUK KOMMUNIKATION
Gesellschaft Fur ProduktPletschmuhlenweg
7
5024 Pulheim
5024
Germany
Voice Phone: 49 (02238) 5 10 51
Description: Holograms, Holographic projects

LAWRENCEBERKELEYLABORATOR"
University of California
Building 80-101
Berkeley
CA94704
USA
Voice Phone: (510) 486 4000
Contact: Malcolm Howells
Description: Industrial & academic hologra-
phy research.

LAZA HOLOGRAMS
47 Alpine Street
Reading
Berkshire
England RG 1 2PY
United Kingdom
Voice Phone: (44) 0734 589 026
Fax Phone: (44) 0734 571974
Contact: Carole Lambert
Description: Specialist mass producer high
quality silver halide holograms. Large or
small quantities. Stock holograms

LAZAPINC
1614 N. Sweetbriar Drive
Claremont
CA91711
USA
Voice Phone: (714) 624-9923
Contact: Richard Cook
Description: Manufacturers of and consultants for NDT and HOE applications. Manufacturers of various types of lasers.

LAZART HOLOORAPHICS.
22 Erina Valley Road
Erina
New South Wales 2250
Australia.
Voice Phone: (61)(043) 676 245
Fax Phone: (61)(043) 652306
Contact: Brett Wilson
Description: Artistic holography; buying & selling holograms.

LAZER WIZARDRY
5650 W. Washington
Suite C-7
Denver
CO 80226
USA
Voice Phone: (303) 297 9930
Fax Phone: (303) 297 9930
Contact: Richard M. Os ada
Description: Wholesale holography. Servicing specialty holographic and other retail stores. One of the largest selections available anywhere, including some industry exclusives. Custom broker, all types of holography-- all markets.

LENOX LASER.
1 Green Glade Court
Phoenix
MD21131
USA
Voice Phone: (301) 592-3106
Fax Phone: (301) 592-3362
Contact: Joseph P. d'Entremont
Description: Laser-systems laboratory specializing in laser drilling, electron beam welding, Edm machining, and water jet. Offers pre-fabricated aperture kits.

LEONARD KURZ GMBH & CO.
Schwabecher Strasse 482
Postfach 1954
Firth
0-8510
Germany
Voice Phone: (49)(0911) 71410
Description: Manufacturer of embossing equipment; broker for hologram embossing.

LESPRODUCTIONSHOLOLAB!
3970, Boulevarde St. Laurent
Montreal
Quebec H2W 1 Y3
Canada
Voice Phone: (514) 849 4325
Contact: Marie-Christiane Mathieu
Description: Artistic holography

LESEBERG,DR.DETLEF
Kamener Str. 172
Lilnen-Beckinghausen
W4670
Germany
Contact: Dr. Detlef Leseberg
Description: Scientific holography research., HOE, computer-generated holography.

LETTERHEAD PRESS INC.
155 North 120th Street
Dept. HM
Wauwatosa
WI 53226
USA.
Voice Phone: (414) 258 1717.
Fax Phone: (414) 2589687
Contact: Mark Mulvaney
Description: Full service trade finisher with 24-hour, 7 -days/week manufacturing. Featuring 19 x 25 inch and 40 inch formats for holographic stamping. Complete projects from print to final bindery assuring singlesource responsibility.

LEVINE, CHRIS
6b Wescott Rd
Kennington
London
England SE17 3QY
United Kingdom
Voice Phone: (44) 071582814
Fax Phone: (44) 0714902693
Contact: Chris Levine
Description: Fine art holograms.

LEXEL LASER, INC.
48503 Milmont Drive
Fremont
CA94538
USA
Voice Phone: (510) 770-0800.
Fax Phone: (510) 651-6598.
Contact: Len Goldfine.
Description: Lexel produces the highest quality Argon and Krypton laser systems. In particular, Lexel specializes in production of single frequency systems which are very stable over a variety of environmental situations.

LIBERATO, PAUL
600 15th St.
Apt. 6
Miami Beach
FL33139
USA
Voice Phone: (305) 672 2685
Fax Phone: (305) 672 0157
Contact: Paul Liberato
Description: Holographic fine artist

LICONIX
3281 Scott Boulevard
Santa Clara
CA95054
USA.
Voice Phone: (408) 496 0300
Fax Phone: (408) 492 1303.
Contact: Terry Erisman
Description: LiCONiX, long the recognized leader in Helium Cadmium laser technology, also suplies semiconductor diode laser systems and a recently introduced line of ion lasers.

LIGHT ENGINEERING
12 New St. Johns
St. Helier
Jersey
Channel Islands
England United Kingdom
Fax Phone: (44)(534) 30614.
Contact: Anthony Hopkins.

LIGHT FANTASTIC PLC
4E/F Gelders Hall Road
Shepshed
Leicestershire
England LE12 9NH
United Kingdom
Voice Phone: (44)(509) 600 220.
Fax Phone: (44)(509) 508 795
Contact: Peter H.L.Woodd;
Description: A fully-integrated holographic business providing the creative and technical services that produce innovative standard and custom-designed holograms of the highest quality. Total service covers embossing and finished product.

LIGHT IMPRESSIONS EUROPE PLC.
5 Mole Business Park 3
Leatherhead
Surrey
England KT22 7BA
United Kingdom
Voice Phone: (44) 0372 386677
Fax Phone: (44) 0372 386548
Contact: John Brown
Description: Branch Office of Light Impressions Inc., Santa Cruz, CA USA.

LIGHT IMPRESSIONS, INC.
149-B Josephine Street
SantaCruz
CA95060
USA.
Voice Phone: (408) 458 1991.
Fax Phone: (408) 458 3338.
Contact: Fred Black
Description: Light Impressions is an integrated, full-service commercial holography company. We produce custom and stock hologram masters and embossallized polyester. Diecutting and hot stamping are also offered.

LIGHT WAVE GALLERY
North Pier
435 East Illinois Street
Chicago
IL 60611
USA.
Voice Phone: (312) 321 1123.
Contact: Shana Wills.
Description: Gallery, retail shop.

LINDA LAW HOLOGRAPHICS
211 Depot Rd
Huntington Station
NY 11746
USA
Voice Phone: (516) 673 3138
Fax Phone: (516) 547 0138
Contact: Linda Law
Description: Educational programs for schools, K-12, on a national basis, teacher training programs & studio workshops for general public. Curates exhibitions of fine art holography as well as producing her own fine art images.

LINE LITE LASER CORPORATION
430 Ferguson Drive
Bldg. 4
Mountain View
CA94043
USA
Voice Phone: (415) 969-4900.
Fax Phone: (415) 969-5480.
Contact: Sid Schreeve
Description: Manufacturer of low and medium-power CW sealed C02 lasers. Also low, medium, high-power, CW and pulsed Nd: YAG lasers; laser gases; power supplies for gas and solid state lasers.

LITTON SYSTEMS CANADA LTD.
25 Cityview Drive
Rexdale
Ontario M9W 5A7
Canada
Voice Phone: (416) 249-1231
Contact: Romuald Pawluczyk
Description: Holographic non-destructive testing; electro optics.

LONE STAR ILLUSIONS
2901 Capital of Texas Highway
#168
Austin
TX 78746
USA
Voice Phone: (512) 3283599
Contact: Alan Lifshen
Description: Description: Gallery and retail shop.

LOS ANGELES
SCHOOL OF HOLOGRAPHY
P.O. Box 851
Woodland Hills
CA 91365
USA
Voice Phone: (818) 7031111
Fax Phone: (818) 7031182
Contact: Jerry Fox
Description: The Los Angeles School of Holography offers a 3 day class. Students learn all phases of holography, and produce both laser viewable transmission and white light viewable holograms in silver halide format

LOUGHBOROUGH UNIV. OF TECH.
Dept. of Physics
Loughborough
Leicestershire
England LE11 3TU
United Kingdom
Voice Phone: (44)(509) 263171.
Fax Phone: (44) 509 219 702
Contact: Professor Nick Phillips.
Description: Embossing masters/shims; Scientific, industrial research. The University and Markem Systems (UK) are participating in a joint venture, Advanced Holographic Laboratories.

LPT LASER PHYSIKTECHNIK GmbH
Waldbahnweg 16
Sauerlach
8029
Germany
Voice Phone: (49) (08104) 1077 + 1
Fax Phone: (49) (08104) 9336
Description: Lasers, used.

LULEA UNIVERSITY OF TECHNOLOGY
Dept. of Mechanical Engineering
S-951 87 Lulea
Sweden
Contact: Nils-Erik Molin
Description: Industrial research; holographic non-destructive testing.

LUMONICS INC.
Company Headquarters
105 Schneider Road
Kanata, Ottawa
Ontario K2K 1 Y3
Canada.
Voice Phone: (613) 592 1460.
Fax Phone: (613) 592 5706
Contact: Dr. Jim Higgins
Description: Lumonics is a manufacturer of high power pulsed ruby lasers for portrait holography and engineering holocameras for NDT. Other products include laser marking and materials processing systems.

LUMONICS LTD
Cos ford Lane
Swift Valley
Rugby, Warwickshire
England CV211QN
United Kingdom.
Voice Phone: (44) 0 788 570321
Fax Phone: (44) 0788579824.
Contact: George Synowiec
Description: Lumonics manufactures pulsed lasers for a range of industrial and scientific applications including pulsed ruby lasers for Holography. Single pulse and multiple pulse units available for commercial, research and NDT applications.

LUND INSTITUTE OF TECH.
Department of Physics
Box 118
Lund
S-221
Sweden
Voice Phone: (46) (046) 107656
Contact: Sven-Goran
Description: Color H-l; holography education; academic research.

LURE
Institut d 'Optique
BP 147
Orsay
Cede x F-91403
France
Voice Phone: (33)(1)(69) 416846.
Contact: D. Joyeux.
Description: Academic research.

MACSHANEHOLOGRAPHY
c/o Laser Arts Programs
512 West Braeside Drive
Arlington Heights
IL60004
USA.
Voice Phone: (708) 398 4983.
Contact: Jim MacShane
Description: Design and manufacturing Sunbows' sculptural, architectural, and gift embossed holographic products, educational programs and artistic holography.

MAGIC LASER (France)
Quartier de L'horloge
4 rue Brantome
Paris
5003
France
Voice Phone: (33)(1) 4274 3578
Fax Phone: (33)(1) 42743357
Contact: Anne-Marie Christakis
Description: Importer and wholesaler of all
holographic products--travelling exhibition.

MAGIC LASER (SPAIN)
General Peron 32
6M
Madrid
78020
Spain
Voice Phone: (34) 1 555
Fax Phone: (34) 1 5565427
Contact: A.M. Christakis
Description: Wholesaler, travelling hologra-
phy exibit

MAGIC LIGHT HOLOGRAFIE GALLERIE
Bahnhofsplatz 2
Munich 2
D-8000
Germany
Voice Phone: (49)(089) 595 981
Description: Artistic holography; Gallery

MAN ENVIRONMENT, INC.
2251 Federal Avenue
Los Angeles
CA90064
USA.
Voice Phone: (310)477 7922.
Contact: Gary Fisher.
Description: Artistic holograms, holography
systems, equipment R&D, I-step rainbow
holography, multiplex.

MARKEM SYSTEMS LTD.
Astor Road
(Eccles New Road)
Salford
Manchester
England M5 2DA
United Kingdom.
Voice Phone: (44)(61) 789 8131.
Fax Phone: (44)(61) 707 5315
Contact: Jane Oliver
Description: "One-Stop-Shop" for embossed
hot stamping foil and laminating film,
including everything from origination to foil
manufacture

MARKETING & PROMOTION CENTER
Sales Promotion Division
1-5, Taito
Taitoku
Tokyo
110
Japan
Voice Phone: 03-3835-5832
Fax Phone: 03-3831-3896
Contact: Mr. Takahide Mikarni
Description: Hologram Manufacturer.

MARTINSSON ELEKTRONIK AB.
Instrumentvagen 16
Box 9060
HAGERSTEN
S-12609
Sweden
Voice Phone: (946)(08) 744 0300
Fax Phone: (46)(08) 744 3403
Contact: Per Skande
Description: Artistic; pulsed portraiture;
equipment & supplies.

MARUBUNCORPORATION
8-1 Nihombashi Odemmacho
Chuo-Ku
Tokyo
103
USA
Voice Phone: (81)(03) 648 8115
Fax Phone: (81)(03) 648 9398
Description: Branch office of Newport Corpo-
ration, Fountain Valley, CA USA

MASSACHUSETTS INSTITUTE
OF TECHNOLOGY
Media Laboratory/Spatial Imaging
20 Ames Street # E15-421
Cambridge
MA02139
USA.
Voice Phone: (617) 2530632.
Fax Phone: (617) 258 6264.
Contact: Linda Conte.
Description: College holography courses;
Computer Generated Holography research

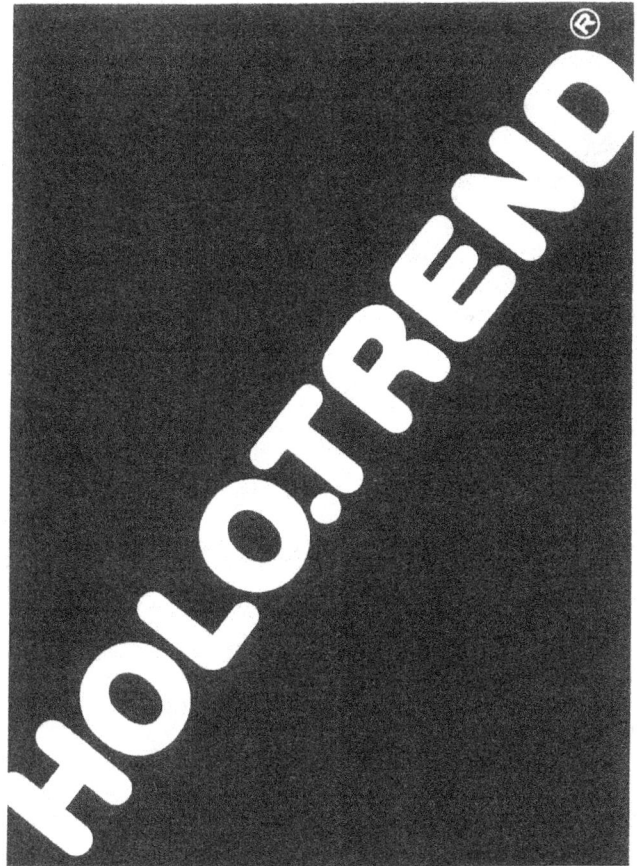

MAZDA MOTOR CORP.
TecluUcal Research Center
P.O.Box 18
Hiroshima
73091
Japan
Voice Phone: 082-282-1111
Fax Phone: 082 252-5343
Contact: Mr. Ichiro Masamori
Description: Holographic Interferometry

MBB-Industrieerzeugnisse Rosenheimer S
trasse
München 80
8000
Germany
Voice Phone: (089) 6 00 00
Description: Laser, medical

MC CORMACK, SHARON
P.O. Box 38
White Salmon
WA98672
USA
Voice Phone: (509) 493 4850
Fax Phone: (509) 493 4830
Contact: Sharon McCormack
Description: Holographic fine artist.

MEDIA ARTS TEACHER ASSOCIATION
4101 Taylor Rd
Jamesville
N.Y. 13078
USA
Voice Phone: (315) 4698574
Contact: Bob Reals
Description: "Holography in the Classroom"
program. Includes workshops for teachers,
exhibitions of student-created holograms, a
six-week summer school for gifted holography
students, curriculum materials.

MEDIA INTERFACE, LTD.
215 Berkeley Place
Brooklyn
NY 11217
USA.
Voice Phone: (718) 3981136.
Fax Phone: (718) 3981136
Contact: Ronald Erikson
Description: Consulting in holographic ap-
plications and display. Design, art direction,
and exhibition concepts developed for clients.
Holography in education and curriculum
consultants.

MELLES GRIOT
1770 Kettering Street
Irvine
CA92714
USA
Voice Phone: (714) 261-5600.

Fax Phone: (714) 261-7589.
Contact: . Sales
Description: Melles Griot is a major manufacturer of off-the-shelf and custom tables and
isolation equipment, laser, lenses, mounting
hardware, positioners, polarizers, coated optics, detectors, collimators and spatial filters.

MELLES GRIOT GmbH
Kleyerstrasse 14
Darmstadt
6100
Germany
Voice Phone: (49) (06151) 86331
Fax Phone: (49) (06151) 8235
Description: Condensers (optics), fiber optics
construction components, laser diodes laser
optics,optical filters ,optical lenses, optical
mirrors, optical parts, bulk, optoelectronic
components, planar optics,planar parallel
optics, prisms

MEREDITH INSTRUMENTS
5035 North 55th Avenue
Suite 5
Glendale
AZ 85301
USA
Voice Phone: (602) 934 9387.
Fax Phone: (602) 934 9482
Contact: Chad Andersen
Description: Specializing in surplus inven-
tories of He-Ne lasers as well as argon and
diode lasers, Meredith Instruments is the
USA's largest laser discount dealer. Free
Catalogue.

METAMORFOSI OLOGRAFIA ITAUA
Via Lecco 6
Milano
20124
Italy
Voice Phone: (39)(2) 204 9943
Fax Phone: (39)(2) 204 1625
Contact: Eva Aprile
Description: Producer of the first and original
HOLOTIME, the interchageable hologram
watch. This year we are producing small
size DCG holograms. Consulting to Italian
customs marketing.

METAPLAST ELECTROCHEMICALS
CORP.
67 Whitson Street
Hempstead
New York 11550
USA
Voice Phone: (516) 766 4281
Fax Phone: (516) 481 7320.
Contact: J. L. Lester.
Description: Manufacturers of conductive
coatings and silver spray for shims used in
embossing

METROLASER
18006 Skypark Circle
Suite 108
Irvine
CA 92714-6428
USA
Voice Phone: (714) 553-0688.
Fax Phone: (714) 553-0495.
Contact: James D. Trolinger
Description: Provides services and research
in holographic measurement. Operates a holo-
graphic non-destructive testing laboratory.

METRO LOGIC GMBH
Dornierstrasse 2
Puchheim 8039
Munich
Germany
Voice Phone: (49) 89 800 5311
Fax Phone: (49) 89 800 5312
Description: see Metrologic, USA.

METROLOGIC INSTRUMENTS, INC.
P.O. Box 1458
Coles Road
@Route42
Blackwood
NJ 08012
USA
Voice Phone: (609) 2288886
Fax Phone: (609) 228 6673 .
Contact: Betty Williams
Description: Retail/industrial: laser bar-code
scanners including portables, compact min-
islots and high-speed projection scanners.
Education: HeNe lasers; optics lab, sandbox
holography lab; optics bench system.

MGD PRODUCTIONS
(George Dyens)
5982, Rue Durocher
Outremont
Quebec H2V 3Y4
Canada
Voice Phone: (514) 278-4593.
Fax Phone: (514) 987-4651
Contact: George Dyens
Description: Produces large format holograms
to be integrated in architectural or multi-
media projects in collaboration with designers
or architects.

MGM CONVERTERS INC
10600 Pioneer Blvd.
Santa Fe Springs
CA90670
USA
Voice Phone: (213) 946 3441
Fax Phone: (213) 944 0524
Contact: Steve Meyer
Description: Quality graphics serving the
holography market

MICRAUDEL.
93 rue Adelshoffen
Schiltigherrn, F-67300
France
Voice Phone: (33) 8883 7576
Fax Phone: (33) 88813293.
Contact: Delepiere
Description: Holographic films for holographic
camera. Holographic camera for non-destruc-
tive testing and interferometric holography.

MIDWEST LASER PRODUCTS
P.O. Box 2187
Bridgeview
II., 60455
USA
Voice Phone: (708) 460 9595
Fax Phone: (708) 430 9280
Contact: Steve Garrett
Description: New and used laser equipment
including: HeNe, Argon, HeCd, NdYAG, vis-
ible diode lasers. Manufactures low-cost HeNe
lasers for lab use. Distributes holography kit,
and related materials.

MILLER, PETER
536 Sumner
SantaCruz
CA95062
USA
Voice Phone: (408) 4291002
Contact: Peter Miller
Description: Holographic Fine Artist.

MIND'S EYE
HOLOGRAPHIC CONSULTANTS
17329 Zola Street
Granada Hills
CA91344
USA
Voice Phone: (818) 360 6023
Contact: Stephen Roth
Description: Marketing consultant

MINISTRY
TRADE
OF INTERNATIONAL
Electrotechnical Laboratory
Optical Information Section
Agency of Industrial Science
Tsukuba Science City
305
Japan
Voice Phone: 0298 58 5625
Fax Phone: 029858 5627
Contact: Dr. Satoshi Ishihara
Description: Research using HOEs

MITSUBISHI HEAVY INDUSTRIES LTD.
Nagasaki Technical Institute
1-1 Akunoura-machi
Nagasaki
850-91

Japan
Contact: M. Murata
Description: Holographic non-destructive
testing; industrial research.

MITUTOYO MEASURING INSTRUMENTS.
18 Essex Road
Paramus
NJ07652
USA
Voice Phone: (201) 368 0525.
Contact: Joe Scriff
Description: Manufacturers of precision
measuring instruments including holographic
linear tracking systems.

MMG MINNAHUETTE MASCHINELLE
GLAS
Wittenstrasse 1
Goslar
3380
Germany
Voice Phone: (49) (05321) 24055
Fax Phone: (49) (05321) 2024
Description: Flat glass, refined, front surface
mirrors,glass components,mirrors, surface-
coated, optical mirrors,surface-coated mirrors

MODERN OPTICS
A Division of V-Tech, Inc
270 East Bonita Avenue
Pomona
CA91767
USA
Voice Phone: (714) 596-7741
Fax Phone: (714) 596-9033
Contact: Irvin Miller

MOELLER WEDEL OPTISCHE WERK
Rosengarten 10
Wedel
2000
Germany
Description: Artistic holography.

MOSCOW PHYSICAL ENGINEERING
INSTITUTE
Kashirskoe Shosse 1
Moscow
115409
Russia
Contact: Alexander Larkin
Description: Artistic hoography; scientific
holography research.

MULTI-PURPOSE HOLOORAMS (MPH).
E
6 rue de l'orme
Paris
75019
France
Voice Phone: (331) 42009866
Fax Phone: (331) 47434744
Contact: Fani Adam or Edwina Alva
Description: Holograms' conception, realization and application is our domain. We undertake all kinds of research in holography, not only scientific (mastering) but also artistic, such as self-illuminated holo-sculptures.

MUNDAY SPATIAL IMAGING.
39 Pyrcroft Road
Chertsey
Surrey
England KT16 9HT
United Kingdom
Voice Phone: (44)(0932) 564 899
Contact: Rob Munday
Description: Specializes in holograms of museum artifacts. Display Reflection and Rainbow holography of CW, Pulsed and Stereogram images. Computer/Video imaging holographic stereogram system. Collection of 200 holograms for exhibition.

MUSEUM FuR HOLOGRAPHIE
& NEUE VISUELLE MEDlEN
Pletschmuhelenweg 7
Pulheim 1
D-5024
Germany
Voice Phone: (49)(02) 233 385 1053
Fax Phone: (49)(02) 238 52158
Contact: Matthias Lauk
Description: The first museum of holography in Europe. Permanent showroom including classical holographic artworks. Guided tours. Workshop program. Consultation and organisation of national and international exhibitions. Holography Consulting.

MUSEUM OF HOLOORAPHY/CHICAGO.11
1134 West Washington Blvd
Chicago
IL60607
USA
Voice Phone: (310) 226 1007
Fax Phone: (312) 829 9636
Contact: Loren Billings.
Description: The Museum of Holography/Chicago is the only museum in the USA devoted exclusively to holography as a art form. regular classes, exibits, R&D and a library available.

MUSèE DE L'HOLOORAPHIE
15 Ii 21 Grand Balcon
Forum des Halles
BP 180,
Paris
75001
France
Voice Phone: (33)(1) 4039968
Fax Phone: (33) 42743357
Contact: Anne-Marie Christakis,
Description: Permanent & travelling exhibition; educational courses.

MU's LASER WORKS
1328 Dunsterville Avenue
Victoria
B.C. V8Z 2Xl
Canada V8Z
Voice Phone: (604) 479-4357
Contact: Ron Meuse
Description: Holographic and 3-D photographic services, Can provide lab rental and technical assistance. Laser light show production and rental.

MWK INDUSTRIES
198 Lewis Court
Corona
CA91720
USA
Voice Phone: (714) 278-0563.
Fax Phone: (714)
Contact: Mike Kenny.
Description: Distributor of HeNe and argon lasers

MYERS PRINTING, INC.
914 East Gier Street
Lansing
MI 48906
USA
Voice Phone: (800) 333 3645
Fax Phone: (517) 482-0550
Contact: Jim Dick
Description: Four years' experience in applying holograms. We have worked with holograms generators and others applicators across the U.S. We look forward to the opportunity to work with additional holography experts.

McCAIN MARKETING.
10962 North Wauwatosa Road
76W
Mequon
WI 53092
USA.
Voice Phone: (414) 242 4023.
Contact: Richard McCain
Description: Act as liason between commercial advertisers (including Fortune 500 companies) and holographers. Educate clients to holography, develop advertising promotions, and educational applications using holography. Recommend professional holographic specialists as needed

McMAHAN ELECTRO-OPTIC
2160 Park Avenue
(Orlando Division)
Winter Park
FL32789
USA
Voice Phone: (407) 645-1000
Fax Phone: (407) 644-9000
Contact: Robert McMahan, Sr.
Description: McMahan Electro-Optics manufactures a laser-based NDT system for testing composite aerospace components and assemblies.

MOLLER GmbH
Am Bleichbach 7
Moosinning
8059
Germany
Voice Phone: (49) (08123) 12 11
Description: Holography Artist

NAIMARK, MICHAEL
216 Filbert St.
San Francisco
CA94133
USA
Voice Phone: (415) 3914817
Contact: Michael Naimark
Description: Fine Art holograms.

NASA MARSHALL SPACE FUGHT
CENTE
Space Sciences Laboratory
ES 73
HlUltsville
AL 35812
USA
Description: Scientific holography

NATIONAL HOLOGRAPHIC STUDIOS
Technology Development Center
Kresson-Gibbsboro Road
Voorhees
NJ08043
USA
Voice Phone: (609) 784-3800.
Fax Phone: (609) 784-2471
Contact: Mike Chappelle.
Description: Artistic holography; fine art limited editions, holographic stereograms; commercial H-l. Custom presentation support.

NATIONAL PHYSICAL LABORATORY
Queens Road
Teddington
Middlesex
England 1Wll OLW
United Kingdom.
Voice Phone: (44)(81) 977 3222.
Contact: D. Robinson.
Description: Scientific and industrial research; holographic non-destructive testing.

NEOVISION PRODUCfIONS
P.O. Box 74277
Los Angeles
CA90004
USA.
Voice Phone: (213) 387 0461
Contact: Bill Hillard
Description: Fine art originals, producing holograms for home and industry, consulting.

NEW CLEAR IMPORTS LTD.
27 Burrard Street
St. Helier
Jersey
Channel Islands
England United Kingdom.
Voice Phone: (44)(534) 30614.
Contact: Anthony Hopkins
Description: Gallery; retail shop.

NEW DIMENSION HOLOGRAPIDCS.
27 Nurses Walk
The Rocks,
Sydney
New South Wales 2000
Australia.
Voice Phone: (61)(2) 2476063.
Fax Phone: (61) 2419 8670.
Contact: Tony Butteriss.
Description: Artistic holography sales; Educational holography consultant.

NEW YORK HALL OF SCIENCE
47-01111thStreet
Corona
NY 11368
USA.
Voice Phone: (718) 699 0005.
Contact: John Driscoll
Description: The New York Hall of Science is New York's only hands-on science and technology museum. Lasers and optics demonstrated daily. Color hologram depicting quantum atom is on display.

NEW YORK HOLOGRAPIDC LABS
P.O.Box 20391
Tompkins Square Station
176 East 3rd Street
New York
NY 10009
USA
Voice Phone: (212) 254 9774
Contact: Daniel Schweitzer or Samuel More
Description: Private Commissions, Fine Art Holograms, Hands-on Tutorials.

NEW YORK INSTITUTE OF TECH.
Center for Optics
100 Glen Cove Avenue
Glen Cove
NY 11542
USA.
Voice Phone: (516) 686 7863.
Contact: Maurie Haliva.
Description: Non-Destructive Testing; Industrial research; and Interferometry.

NEWPORT (ASIAN OFFICE)
Kyokuto Boeki Kaisha
7th Floor
New Otemachi Bldg.
2-1,2-Chome,
Otemachi Chiyoda-ku, Tokyo 100-
Japan.
Description: Branch office of Newport Corp.,
Fountain Valley CA, USA. Newport

NEWPORT CORPORATION
(Main Office)
18235 Mt. Baldy Circle
P.O.Box 8020
Fountain Valley
CA 92708-8020
USA.
Voice Phone: (714) 9639811
Fax Phone: (714) 963 2015
Contact: Jim Doty
Description: Designer and manufacturer of laserlholographic systems, E/O componenlSV optics, spatial filters, optical & bearnsteering instruments, magnetic bases, fi ber optic components, vibration isolation systems. an holographic recording materials.

NEWPORT GMBH
European Headquarters
Bleichstrasse 26
Darmstadt
D-6100
Germany
Voice Phone: (49) 0615126116.
Fax Phone: (49) 0615122639.
Description: Branch office of Newport Corporation, Fountain Valley, CA. USA.

NOT LISTED? PLEASE FAX US AT (1) 510 841 2695 AND WE WILL ADD YOU TO OUR MAILING LIST FOR THE NEXT EDITION ...

NIHON UNIVERSITY
Department of Electronic Engineering
College of Science and Technolo
7-24-1 Narashinodai
Funabashi -shi
Chiba 274
Japan
Voice Phone: 0474 69 5391
Fax Phone: 0474 67 9683
Contact: Dr. Hiroshi Yoshikawa
Description: Research NDT

NIPPON POLAROID K.K.
Marketing Division.
Mori Bldg. No.30
3-2-2 Toranomon
Tokyo (Minato-ku)
105
Japan
Voice Phone: 03 3438 8874
Fax Phone: 03 3433 3537
Contact: Mr. Hiroyuki Hagura

NIPPONDENSO CO., LTD.
System Development Engineering
1-1 Showa-cho
Kariya-shi
Aichi-ken
448
Japan
Voice Phone: 056625-6924
Contact: Mr.Hiroshi Ando
Description: Manufacture HeadsUp Display.
Also Mr. Toru Mizuno, Mr. Tatsuya Fujita, or
Mr. Shinji Nanba.

NIPPONDENSO Co., Ltd.
System Develop. Engineering Dept
1-1 Showa-cho
Kariya-slu
Aichi-ken
448
Japan
Voice Phone: 0566 25-6924
Contact: Mr. Toru Mizuno
Description: Hologram manufacturer.

NISSAN MOTOR.
Central Research Lab
Natsushima machi
Yokosuka
237

Japan
Voice Phone: 0468 62-5182
Fax Phone: 046 654183
Contact: Mr. Shigeru Okabayashi
Description: Hologram manufacturer headup display

NORGES TEKNISKE HOGSKOLE.
Institute for Almen Fysikk
Sem Saelandsv 7
N-7034
Trondheim-NTH
Norway
Voice Phone: (47)(07) 5
Fax Phone: (47)(07) 592886
Description: Holographic Non--destructive testing

NORLAND PRODUCTS, INC.
695 Joyce Kilmer Avenue
P.O. Box 145
North Brunswick
NJ08902
USA
Voice Phone: (908) 545 7828.
Fax Phone: (908) 545 9542.
Contact: Jean Spalding.
Description: Manufacture and distribute holographic photochemistry.

NORTH AMERICAN HOLOGRAPHICS
P.O. Box 451
103 East Scranton Avenue
Lake Bluff
IL60044
USA.
Voice Phone: (708) 234 4244.
Contact: Gary Lawrence.
Description: Holographic portraits; Holography marketing & applications; Trade shows & exhibits.

NORTH DANCER HOLOGRAPHIC
3654 North Dancer Road
Dexter
MI48130
USA
Voice Phone: (313) 426-0452
Contact: Charles Lysogorski
Description: Holographic non-destructive testing, production display, consulting, artistic limited editions.

NORTHERN ILLINOIS UNIVERSITY.
Department of Physics
DeKalb
IL 60115
USA
Voice Phone: (815) 7531772
Contact: Thomas Rossing.
Description: Scientific holography research

NORTHERN LIGHTWORKS
2148 North 86th Street
Seattle
WA98103
USA
Voice Phone: (206) 526 5752.
Contact: Edward Aites
Description: Holographic fine art imagery IT. transmission and reflection formats. Unique and limited-edition artwork.

NORTHWESTERN UNIVERSITY
Dept of Biomedical Engineering
Evanston
IL60208
USA.
Voice Phone: (708) 491 2946
Fax Phone: (708) 491 4133
Contact: Hans Bjelkhagen
Description: Artistic, commercial, interferometry, scientific, medical, and industrial research, portraits, fiberoptics, pulsed laser holographic non-destructive testing, educational holography.

NORTHWESTERN UNIVERSITY
Fiberoptic Laboratory
Dept of Electrical Engineering,
Technological Institute
Evanston
IL60208
USA
Voice Phone: (708) 4914427
Contact: Rudy Haidle
Description: Optical processing, Forensic holography.

NOVA TOR RESEARCH CENTER.PI.
Lesi Ukrainki 1
Kiev
196252196
Russia
Voice Phone: (7)(296)8048
Contact: Georgi Dovgalenko
Description: Holography applications to medicine.

NUMAZU COLLEGE OF TECHNOLOGY
Department of Mechanical Enginee
3600 Ooka
Numazu-city
Shizuoka 410
Japan
Voice Phone: 055921 2700
Contact: Dr. Koji Lkegami
Description: Holographic research.

O/E RESEARCH
7227 Eastwood Street
Phildelphia
PA 19149
USA

Voice Phone: (215) 3315067
Contact: Len Stockler
Description: Technical and educational consulting. Holography workshops, resource center. Conceptual design and production of display holograms. H-l and H-2 mastering. Custom optical table components. Touring programs, exhibitsd workshops.

ODHNER HOLOGRAPIDCS
P.O. Box 56-8574
Orlando
FL32803
USA
Voice Phone: (407) 856 7665
Fax Phone: (407) 856 7665
Contact: Jeff Odhner
Description: Exclusive distributor of the Stabilock IT inch fringe stabilizer (used to make brighter holograms), manufacture of custom holograms (trans/refl.). Specializing in HOE arrays (to 8" x 10") on silver halide.

OMNICHROME
Quality Lasers and E/O Systems
13580 Fifth Street
Chino
CA 91710
USA
Voice Phone: (714) 627-1594
Fax Phone: (714) 591-8340
Contact: Kevin Rankin
Description: Manufacturer of Argon, Krypton and HeCd lasers ranging in wavelength from Red to UV. Our rugged, hands-off operation lasers find applications in semiconductors, optical disk mastering, medicine and holography.

ONTARIO COLLEGE OF ART
100 McCaul Street
Toronto
Ontario M9W 1 WI
Canada.
Voice Phone: (416) 977-5311
Contact: Michael Page
Description: Holography courses

ONTARIO HYDRO. RESEARCH DIVISION
KR 128
Toronto
Ontario
M8Z 5S4 Canada.
Voice Phone: (416) 231-4111
Contact: David Mader
Description: Holography research; holographic non-destructive testing.

ONTARIO SCIENCE CENTRE
770 Don Mills Road
Don Mills
Ontario M3C 1 T3
Canada.

Voice Phone: (416) 429 4100
Fax Phone: (416) 429-5961
Contact: Mr. David Szanto
Description: Gallery; courses in holography

OP-GRAPHICS (HOLOGRAPHY) LTD.
Unit 4
Technorth
7 Harrogate Road
Leeds
England LS7 3NB
United Kingdom.
Voice Phone: (44)(0532) 628 687
Fax Phone: (44) (0532) 374 182,
Contact: Valerie Love
Description: Manufacturer of display holograms. Large selection of stock images in variety of formats and sizes. Commissioned work undertaken. Copying work for holographers undertaken.

OPTICAL IMAGES
637 South Vinewood Street
Escondido
CA92029
USA.
Voice Phone: (619) 7460976
Fax Phone: (619) 746 6141
Contact: Donald C. Broadbent
Description: Optical Images, formerly Broadbent Development Lab, is an independent, privately owned holographic facility producing HOE's and display holograms in various recording materials. Donald Broadbent has 24 years experience in holography.

OPTICAL LABORATORY
3, Rue de Universite
Strasbourg,
67000
France
Description: Artistic holography.

OPTICAL SURFACES LTD.
Godstone Road
Kenley
Surrey
England
CR2 5AA United Kingdom
Voice Phone: (44)(81) 668 612
Fax Phone: (44) 81 6607743
Contact: John Mathers
Description: Manufacturer of precision optics

OPTICAL WORKS LTD.
Ealing Science Centre
Treloggan Lane,
Newquay
Cornwall
England TR7 IHX
United Kingdom
Voice Phone: (44) 0637-877222
Fax Phone: (44) 0637-877211
Contact: E.O. Frisk
Description: Make optical components and lenses.

OPTICS PLUS INC
1369 East Edinger Avenue
Santa Ana
CA92705
USA
Voice Phone: (714) 972 1948
Fax Phone: (714) 835-6510
Contact: Allison Frost
Description: Manufacture optics; precision tool mounts (including lens and mechanical mounts).

OPTILAS B.V.
P.O. Box 222
2400 AE Alphen
NDRijn
Netherlands
Voice Phone: (31) 0172031234
Fax Phone: (31) 01720 43414
Contact: A. Kooi
Description: Sales/service/engineering of electro-optical and vacuum related products.

OPTISCHE FENOMENEN
Nederlandse Stichting Voor Waarn
Warenarburg 44
NL2907 CL
Capelle aid Ijssel
Netherlands
Contact: Jan M. Broeders
Description: Monthly Newsletter - subsidary of Dutch Foundation of Preception & Holography

ORIEL CORPORATION
250 Long Beach Boulevard
Stratford
CT06497
USA
Voice Phone: (203) 377-8282
Contact: Scott Heidemann
Description: Holography artist

ORIEL SCIENTIFIC LTD.
(Division of Oriel Corporation)
.P.O. Box 31
Leatherhead
Surrey
England KT22 7AU

United Kingdom
Voice Phone: (44) (0372) 378822
Description: Artistic holography.

OXFORD HOLOGRAPHICS.
71 High Street
Oxford
England OXI 4BA
United Kingdom
Voice Phone: (44)0865250505
Fax Phone: (44) 0865 790 353
Contact: Nick Cooper
Description: Oxford Holographics has both a
very well established retail and an expanding
distribution operation, focusing on silver
halide & innovative embossed products.

OXFORD UNIVERSITY
Dept. of Engineering Science
Parks Road
Oxford
England OXI 3PJ
United Kingdom
Voice Phone: (44) (0865) 273 805
Contact: D. J. Cooke.
Description: Holography education; Industrial
research.

PACIFIC HOLOGRAPHICS, INC.
3003 Brandon Circle
Carlsbad
CA92oo8
USA
Voice Phone: (619) 434 4466
Fax Phone: (619) 4344466
Contact: Eric Van Hamersveld.
Description: Specialize in complete stereo-
gram production, mastering & final product.

PALCO'S HOLOGRAM WORLD
(Hologram World Distributors)
1884 Berkshire Lane North
Plymouth
MN55442
USA
Voice Phone: (612) 559-5539
Fax Phone: (612) 559 2286
Contact: Jim Paletz
Description: One of the largest wholesale
distributors of holographic novelties. We
represent over 50 holographic manufacturers.
Free catalogue.

PARKER, JULIE WALKER
c/o Diffraction Ltd.
P.O. Box 1115
Waitsfield
VT05673
USA
Contact: Julie Walker Parker
Description: Holographic Fine Artist; multi-
color reflection holograms.

PENNSYLVANIA PULP & PAPER CO.
Station Mews #7
7804 Montgomery Avenue
Elkins Park
PA 19117
USA
Voice Phone: (215) 635 0801
Fax Phone: (215) 635 6114
Contact: Brian Monaghan
Description: Manufactures Prismatic Illusions

PENNSYLVANIA STATE UNIVERSITY
Applied Research Laboratory
P.O.Box 30
State College
PA 16804
USA
Contact: C.S. Vikram
Description: Scientific research

PENNSYLVANIASTATE UNIVERSITY
Dept. of Electrical Engineering
121 East University Park
PA 16802
USA
Contact: Francis Yu
Description: Color holography

PHANTASTICA
Hohenweg 14
5760 Arnsberg 1
Germany
Voice Phone: (49) 2932-81917
Fax Phone: (49) 2932-29441
Contact: Gerd M. Albrecht
Description: Makers and distributors of
articles related to embossed holograms and
diffraction foil, including earrings and other
jewelry, badges, pens, mobiles. Main focus is
street, crafts and Christmas markets.

PHASE-R CORPORATION
Box G-2
Old Bay Road
New Durham
NH 03855
USA
Voice Phone: (603) 859 3800
Description: We manufacture laser equip-
ment. Call for more information.

PHOENIX HOLOGRAMS
c/o Elmar Spreer
Traubenstr. 41
Apen
D-2913
Germany
Voice Phone: . (49)044 89 15198
Contact: Elmar Spreer
Description: Phoenix Holograms specializes in
finding individual solutions for sophisticated prob-
lems. We do silver halide holograms up to 1 m2
for: architecture, interior decoration, promotion/
corporate design, education, exhibitions, portraits.

PHOTOGRAPHERS FORMULARY, INC.
P.O. Box 950
Condon
MT59826
USA
Voice Phone: 800 922 5255
Fax Phone: (406) 754 2896
Contact: Bud Wilson
Description: Photochemistry for holography

PHOTON LEAGUE OF HOLOGRAPHERS.
UnitB
Toronto
Ontario
M6J lA 7 Canada
Voice Phone: . (416) 531 7087
Contact: Nicola Woods
Description: Artist run non-profit holography
studio. Technical workshops throughout the
year.

PHOTONICS DIRECTORY
Laurin Publishing Co. Inc.
P.O. Box 4949
Berkshire Common,
Pittsfield
MA01202
USA
Voice Phone: (413) 4990514
Description: Publishers of information on
Optical components; Optics, Electro-Optics,
Lasers, and Imaging technology.

PHOTONICS SYSTEMS LABORATORY
7 Rue de L'universite.
Strasbourg,
67000
France
Voice Phone: 88355150.
Contact: P. Meyrueis
Description: Education, NDT

PHYSICAL OPTICS CORPORATION.
2545 W 237th Street, Suite B
Torrance
CA90505
USA
Voice Phone: (213) 530-1416
Fax Phone: (213) 530-4577
Contact: Kevin Rankin
Description: Phototonics-based high tech
company involved in research, development
and manufacture of holographic optical ele-
ments, including reflection and transmission
HOE's, diffraction gratings and display
holograms.

PHYSICS INSTITUTE
Latvian Academy of Sciences
Riga-Salaspils
229021
Latvia
Voice Phone: 007-0132/947642
Contact: Dr. Kurt Shvarts
Description: Scientific research on recording
materials

PHYSIK INSTRUMENTE (PI) GmbH & C
Siemensstr 13-15
7517 Waldbronn
7517
Germany
Voice Phone: (49) (7243) 6 04-0
Description: Holography, laser, optical instruments, miscellaneous, oscillation insuators,
vibrating dampers

PILKINGTON OPTRONICS
Glascoed Road
St. Asaph
Clwyd, North Wales
England
LL17 011 United Kingdom
Voice Phone: (44) 745588344
Fax Phone: (44) 745 584 258
Contact: Andrew Hurst
Description: Manufacturer of DCG/photopolymer

PINK, PATTY
P.O. Box 24-153
San Francisco
CA94124
USA
Voice Phone: (415) 822 7123
Contact: Patty Pink
Description: Fine arts holography. High quality
glass plate transmission and reflection holograms.
Classes for children and adults. Technical holog-
raphy writing and editing. Laser news editor.

PLANET 3-D
201 Silver Fox Lane
Downingtown
PA 19335
USA
Voice Phone: (215) 8736192
Contact: Rich Cossa
Description: Marketing of holography.

POINT OF VIEW DIMENSIONS, LTD.
45-2903 River Drive South
Jersey City
NJ07310
USA
Voice Phone: (201) 626 8844
Contact: Neal Lubetsky
Description: Point of View Dimensions spe-
cializes in the conceptualization, design, and
execution of holography (all formats and all
sizes) for exhibitionsade shows, premiums,
points of sale, brochures, and annual reports.

POINT SOURCE PRODUCTIONS
P.O. Box 55
Boulder Creek
CA95006
USA
Voice Phone: (408) 3383438
Contact: R.A. Hess
Description: An independant recording studio

POLAROID CORPORATION
Holography Business Group
730 Main Street - 1A
Cambridge
MA02139
USA
Voice Phone: (617) 577-4307
Fax Phone: (617) 577 5434
Contact: Tim Morgan
Description: Polaroid manufactures high
quality photopolymer holograms for high
volume applications. Both custom and stock
holograms are available. Polaroid's high
quality & dependability come from 20 years
of experience in holography.

POLYMER HOLOGRAPHICS, INC.
Old Mingus Art Center
Gymnasium Bldg.
P.O. Box 847
Jerome
AZ 86341
USA
Voice Phone: (602) 634-8823
Fax Phone: (602) 634-8823.
Contact: Robert C. Lauerman
Description: Manufactures high quality
photopolymer holograms using DuPont
OmniDex" photopolymer. Stock and custom
holograms. Labels for printing, packaging,
advertising. Replication service. HOE's.
Transmission and reflection holograms; de-
sign/production of advertising pieces.

POLYMER IMAGE
21411 N. 11th Avenue, Suite #4
Phoenix AZ 85027
USA
Voice Phone: (602) 7804882
Fax Phone: (602) 7804882
Contact: Dan Norton
Description: Full-service holographic lab,
including mastering and production of custom
images. Products include "Whole-FX" hologram mugs, made with polymer film which
are heat-, UV,- water-resistant, dishwasher/
microwave-safe.

PORTS ON, INC. (LASER IMAGES)
9201 Quivira
Overland Park
KS 66215
USA.
Voice Phone: (913) 492 7010
Fax Phone: (913) 492 7099
Contact: Steve Larson
Description: Manufacturers of stock and
custom holograms and holographic products;
total in-house production capabilities in
dichromate, silver halide, and photo-resist.
Stock products include jewelry, watches,
calculators, and framed art.

PRO DESIGN
Mario Liedtke
Pixelerstrasse 34A
4840 Rheda-Wiedenbrtick
4840
Germany
Voice Phone: (49) 524246051
Fax Phone: (49) 524242012
Description: Advertising holograms, Holo-
graphic projects

RAINBOW SYMPHONY INC.
6860 Canby Avenue
#120
Reseda
CA91335
USA
Voice Phone: (818) 708-8400
FAX: (818) 708 8470
Description: Manufacturers of uniquely
designed holographic and diffraction products
for the guift, novelty, advertising, speciality,
premium incentive, souvenir and museum
markets.

RALCON
Box 142
8501 South /400 West
Paradise
UT 84328
USA
Voice Phone: (801) 245 4623
Fax Phone: (801) 245 6672
Contact: Richard Rallison
Description: Design, development and fabrica-
tion of volume holographic optical elements,
(HOEs) including gratings, scanners, multifo-
cus devices, heads up and down displays and
notch filters formed in dichromated gelatin or
photopolymer

RALPH CULLEN HOLOGRAPHICS
c/o UK Optical Supply
84 Wirnborne Road West
Wirnborne
Dorset
England BH21 2DP
United Kingdom
Voice Phone: (44)(202) 886 831
Fax Phone: (44)(202) 742236
Contact: Ralph Cullen
Description: A Consultancy-Design Service
which in association with UK Optical Sup-
plies (Manufacturing) provide customized
holographic optical components. Advice on
component selection and laboratory/studios
designed to any budget is available.

RANDAZZO,DEAN
2022 West Crystal
Chicago
IL 60622
USA
Contact: Dean Randazzo
Description: Holography artist

RANDY JAMES HOLOGRAPHY
503 Caledonia Street
Santa Cruz
CA95062
USA
Voice Phone: (408) 458 4213
Description: Commercial and fine art holography since 1974. Extensive background in all
forms of display holography: design, mastering, and production. Custom quotes, stock
price list available.

RAVEN HOLOGRAPHICS LTD.
Old Saw Mills
Nyewood
Near Petersfield,
Hampshire
England GU3 15HX
United Kingdom
Voice Phone: 0730-821612
Fax Phone: 0730-821260
Contact: Stuart Ainslie-Brown
Description: Fine art holograms

RECONNAISSANCE, LTD.
1 Erica Court
Wych Hill Place
Woking Surrey
England AU22 OJB
United Kingdom
Voice Phone: (44) 483740689
Fax Phone: (44)483740689.
Contact: Ian Lancaster.
Description: The leading international consul-
tancy for market, industry information and
analysis. Publisher of Holography News and
secretariat to the International Hologram
Manufacturers' Association. All clients stud-
ies are fully confidential.

RED BEAM, INC.
9011 Skyline Blvd.
Oakland
CA 94611
USA
Voice Phone: (510) 482-3309
Fax Phone: (510) 482-1214
Contact: Lon Moore
Description: Specializes in the design and
production of master (HI) holograms for
mass production. Produces his own line of
holograms. Clients include Activision, AT&T,
NFL(superbowl) and Polaroid.

REEL IMAGE
p.o. Box 566
Pacifica
CA94044
USA
Voice Phone: (415) 355 8897
Fax Phone: (415) 355 5427
Contact: Roy Bradshaw
Description: Fine art holograms, incorpora-
tion of patented holographic designs into
fishing tackle and fishing lures.

REGAL PRESS INC
Holographics Division
129 Guild Street
Norwood
MA02062
USA
Voice Phone: (617) 7693900
Contact: William Duffey
Description: Holographic embossing, applica-
tion; Artistic holography.

REYNOLDS METALS CO.
Flexible Packaging Division
6603 West Broad Street
Richmond
VA 23230
USA
Voice Phone: (804) 2813969
Contact: Rich Patterson
Description: Embossed holography on packag-
ing materials.

RICHARD BRUCK HOLOGRAPHY.
3312 West Belle Plaine
#2
Chicago
IL 60618
USA.
Voice Phone: (312) 267 9288.
Fax Phone: (312) 267 9288
Contact: Richard Bruck.
Description: A shop producing quality lowvol-
ume runs. Custom and commercial work. Fine
art originals. Experienced installation and
consultation

RICHMOND HOLOORAPHIC STUDIOS
6 Marlborough Road
Richmond
Surrey
England lWI0 6JR
United Kingdom
Voice Phone: (44)(81) 940 5525
Fax Phone: (44)(81) 948 6214.
Contact: Edwina Orr
Description: Holograms up to 1 sq. meter.
Custom, stock, R&D, technology transfer.

RICHTER ENTERPRISES
640 19th Street
Manhattan Beach
CA 90266-2509
USA
Voice Phone: (213) 546-5107
Fax Phone: (213) 545-6757
Contact: Thomas A. Richter
Description: International sales representa-
tive and distributor of optical and electronic
components, CCD's, CCD cameras, Bragg
cells, grating and subsystems. Call for com-
plete information on purchasing or selling
your requirements.

RISO LABORATORY INC.
307-3-2-3 Hisamoto
Takatsu-Ku
Kawasaki-Shi
Kanagawa
213
Japan
Voice Phone: (81) 44 813 1292
Fax Phone: (81) 44 813 1293
Contact: Mr. Takao Kawahara
Description: Hologram manufacturing.

ROBERT SHERWOOD HOLO DESIGN
400 West Erie Street
Suite #405
Chicago
IL 60610
USA
Voice Phone: (312) 944 3200
Contact: Robert Sherwood
Description: Our designers, holographers, and
account service personnel provide you with
the highest quality standards this new and
exciting technology can offer.

ROCHESTER INST. OF TECHNOLOGY
One Lomb Memorial Drive
P.O. Box 9887
Rochester NY 14623
USA
Voice Phone: (716) 4752770
Fax Phone: (716) 475-5804
Contact: Arnold Lungershausen.
Description: RIT offers holography instruction
as part of its Imaging and Photographic Tech-
nology program. Practical applications are
stressed. The faculty do industrial consulting
and offer occasional workshops.

ROCHESTER PHOTONICS CORPORTION.
800' Connor Road
Fairport
NY 14450
USA
Voice Phone: (716) 377-7990
Fax Phone: (716) 377-7913
Contact: G.M. Morris
Description: Designer and manufacturer of
diffractive optical elements for use in preci-
sion electro-optical systems. Supplier of opti-
cal system design services for military and
commercial uses.

ROFIN-SINAR LASER GmbH
Berzeliusstrasse 87
Hamburg 74
2000
Germany
Voice Phone: 49 (040) 73 34 01-0
Fax Phone: 49 (040) 73 3401
Description: C02 lasers for materials process-
ing, Laser components, Laser processing
devices and machines

ROLLS-ROYCE PLCAdvanced
Research Laboratory
P.O.Box 31
Derby
England
DE28BJ
United Kingdom
Voice Phone: (332) 242424
Fax Phone: (332) 249936.
Contact: Ric Parker
Description: NDT for aircraft engines.

ROSS BOOKS
P.O. Box 4340
Berkeley
CA94704
USA
Voice Phone: (510) 8412474
Fax Phone: (510) 8412695.
Contact: Brian Kluepfel
Description: Publisher of Holography Market-
place and Holography Handbook

ROTTENKOLBER
GmbH
Henschelring 15
Kirchheim
8011
Germany
HOLO-SYSTEM
Voice Phone: (49) 89 9 03 00 21
Description: Holographic measuring instru-
ments, Modular irnageprocessing systems,
Nondestructive testing equipment and systems.

ROWLAND INSTITUTE FOR SCIENCE
100 Cambridge Parkway
Cambridge
MA02142
USA.

Voice Phone: (617) 4974657
Contact: Jean-Marc Fournier
Description: Scientific holography research.

ROY,JEAN
4071 Clark Street
Montreal
Quebec H2W lXl
Canada
Voice Phone: (514) 523 2337
Contact: Jean Roy
Description: Holographic Fine Artist.

ROYAL COLLEGE OF ART.
Holography Unit
Darwin Building,
Kensington Gore
London
England SW7 2EU
United Kingdom
Voice Phone: (44)(01) 5845020.
Contact: Rod Murray.
Description: MA (RCA) Holography; twoyear
post-graduate course. Explores creative
holography in both fine art and designrelated
fields. Facilities include Pulse, Krypton,
HeNe, Stereogram and Computer Graphics.

ROYAL INSTITUTE OF TECHNOLOGY
Department of Industrial Metrolo
Stockholm
Sweden 10044
Sweden
Voice Phone: (46)(08) 790 7823
Fax Phone: (46)(08) 790 8219
Contact: Lennart Svennson
Description: Holography education. Holo-
graphic Non-Destructive testing.

ROYAL PHOTOORAPHIC SOCIETY
Salisbury College of Art
Southampton Road
Salisbury
Wiltshire
England SPI 2PP
United Kingdom
Voice Phone: (44)(0722) 23711
Contact: Mr. Pitt
Description: Artistic holography; holography
education.

RUEY-TUNG, MISS. HUNG
A 202
Chigasati-coat
Nango 6-7-12
Chigasaki-shi
Kanagawa 253
Japan
Voice Phone: 0467-85-7750
Contact: Hung Ruey-Tung
Description: Holographic Fine Artist.

RUTHERFORD & APPLETON LABS.
Chilton
Didcot
Oxon
England OXll OQX
United Kingdom
Voice Phone: (44)(0235) 821900.
Contact: Robert Sekulin.
Description: Particle measurements; Holographic non-destructive testing.

SAAB-SCANIA
S-581
88 Linkoping
Sweden
Voice Phone: (46)(013) 129 020
Contact: Sven Malmqvist
Description: Holographic non-destructive testing; scientific holography research.

SAGINAW VALLEY STATE UNIVERSITY
2250 Pierce Road
University Center
MI 48710-0001
USA
Voice Phone: (517) 790-4000
Fax Phone: (517) 790-2717.
Contact: Hsuan Chen
Description: Course instruction on holography; research includes HOEs, multiplex and rainbow holography.

SANDIA NATIONAL LABORATORIES
Combustion Research Facility
Livermore
CA94550
USA
Voice Phone: (510) 294 3000
Contact: Donald Sweeney
Description: Scientific holography research;

SAXB Y, GRAHAM
Wolverhampton Polytechnic
54 Strafford Street
Wolverhampton, West Midlands
England WVl INJ
United Kingdom
Description: Research scientist; author of "Practical Holography"

SCHAFFNER GALLERY
(Newport Holograms)
3333 Bear St.
Suite 202
Costa Mesa
CA92626
USA
Voice Phone: (714) 966 9208
Fax Phone: (714) 966 9209
Contact: ; David Schaffner
Description: The premier gallery of holographic art.

SCHOOL OF HOLOGRAPHY
Museum of Holography/Chicago
1134 W. Washington Blvd.
Chicago
IL60607
USA
Voice Phone: (312) 2261007
Fax Phone: (312) 8299636
Contact: Loren Billings
Description: See Museum of Holography/ Chicago

SCIENCE & MECHANICS INSTRUMENTS
605 East 59th Street
Dept. H-4
Brooklyn
NY 11234
USA
Voice Phone: (718) 5313381
Description: Manufacture and sell light meters and shutters.

SCIENCE KIT & BOREAL LABORATORIE
777 East Park Drive
Tonawanda
NY 14150-6784
USA
Voice Phone: (716) 874 6020
Fax Phone: (716) 8749572
Description: Suppliers and mail-order cataloguers of holography educational materials including holography kits, books and more.

SCIENTIFIC COUNCIL ON EXHIBmON
, Academy of Sciences
30, Varilov Street
Moscow
Russia
Voice Phone: (7)(135) 6
Contact: Larisa Nekrasova
Description: Traveling exibits including holography

SEMICON AUSTRIA
Morellenfeldgasse 41
Graz
A-80lO
Austria
Voice Phone: (0316) 38 25 41
Fax Phone: (0316) 38 2403
Description: Collection of Russian art holograms for sale.

SHANDONG ACADEMY OF SCIENCES
Keyuan Road
Jinan Shandong
Jinan Shandong 250014
China
Voice Phone: 615102316
Contact: Prof. Zhu De Shun
Description: Laser & holography exibit.

SHARP CORP.
Tokyo Research Laboratories
Research Dept.2
27-1, Kashiwa
Kashiwa
227
Japan
Voice Phone: 0471 346166
Fax Phone: 047134 6119
Contact: Mr. Shunichi Sato
Description: Research

SHIPLEY CHEMICAL CO.
1457 McArthur
Whitehall
PA 18052
USA
Voice Phone: (215) 820 9777
Contact: Stu Price
Description: Manufacture holographic photochemistry, film, plates and more.

SIEMENS (USA)
186 Wood Avenue South
Iseling
NJ08830
USA
Voice Phone: (201) 3213400
Fax Phone: (812) 422 2339
Contact: Marketing department
Description: Manufacturer of lasers and components.

SIEMENS LTD.
Siemens House
Windmill Road
Sunbury-on-Thames
Middlesex
England TW16 7HS
United Kingdom
Description: Manufacturer of lasers and components.

SILLCOCKS PLASTICS INTERNATIONAL
310 Snyder Avenue
P.O. Box 605
Berkeley Heights
NJ07922
USA
Voice Phone: (908) 665 0300
Fax Phone: (908) 665 9254
Contact: Tom Palmentieri
Description: Producer of flat plastic products, printed or unprinted, which can feature hot-stamped holograms. Products include credit cards, promotional cards and other custom specialties and POP products.

SMITH & McKAY PRINTING CO. INC.
96 North Almaden Boulevard
San Jose
CA 95110-2490
USA
Voice Phone: (408) 2928901
Fax Phone: (408) 292 0417
Contact: Dave McKay
Description: Parent Company: Holographic
Impressions. Hot-stamp foil holograms onto
paper products. Dimensional printing and
fine lithography. Assist coordination of print-
ing projects featuring embossed holography.
Hold holography seminars for graphic design-
ers. Parent company: Holographic Impres-
sions.

SOCIETA OLOGRAFICA IT AUA
(SOl)
Via degli Eugenii 23
Roma
00178
Italy
Voice Phone: (39)(6) 7180976
Fax Phone: (39) (6) 7185172
Contact: Luigi Attardi
Description: Produces artistic/commercial
holograms of any type.

SOCIETY FOR PHOTO-OPTICAL
INSTRUMENTATION ENGINEERS
(S.P.I.E.).
P.O. Box 10
Bellingham
WA98227
USA
Voice Phone: (206) 676 3290.
Contact: Holography Division
Description: Technical publications on holog-
raphy.

SOPHIA UNIVERSITV
Faculty of Science and Technology
7-1, Kioi-cho
Chiyoda-ku
Tokyo
102
Japan
Fax Phone: 03 32383341
Contact: Ms. Kazue Ishikawa
Description: Holography research.

SOUTHERN INDIANA HOLOGRAPHICS
6841 Newburgh Rd
Evansville
IN 47715
USA
Voice Phone: (812) 474-0604.
Contact: Larry Johann
Description: Holographic Fine Artist.

SOVISKUSSTVO v/o MEZHDUNARONAYA
Art holography department
141120 Tryazino
Moscow region
Russia
Voice Phone: (7) 238 4600
Fax Phone: (7) 230 2117
Contact: German B. Avksentjev
Description: Our firm offers: DCG pendents;
lamps with holograms; art reflection holo-
grams (size 102x127 mm, and 180x240mm.).

SPACE AGE DESIGNS INC.
P.O. Box 72
Carversville
PA 18913
USA
Voice Phone: (215) 2978490.
Contact: Valli Rothaus
Description: Conceive, design, manufacture
products for markets which vary--consulting
services available.

SPECAC LTD
6A River House
Lagoon Road, St. Mary Cray
Orpington
Kent
England BR5 3QX
United Kingdom
Voice Phone: (44) 689 873134
Description: Artistic holography.

SPECTRA-PHYSICS INC.
Laser Products Division
1250 West Middlefield Road
Box 7013
Mountain View
CA 94039-7013
USA
Voice Phone: (415) 961 2550
Fax Phone: (415) 969 4084
Contact: Steve Guggenheimer.
Description: World's largest supplier of CW
and pulsed gas and solid state laser systems,
including a comprehensive optical accessories
line and a worldwide customer service net-
work.

SPECTRATEK CORPORATION
1510 Cotner Avenue
Los Angeles
CA 90025
USA
Voice Phone: (310) 473-4966
Fax Phone: (310) 477-6710
Contact: Mike Wanlass
Description: Specialists in high-resolution,
low-cost holograms. We have delivered hun-
dreds of millions of holograms.

SPECTROGON (USA)
550 County Avenue
Secaucus
NJ07904
USA
Voice Phone: (201) 867 4888
Fax Phone: (201) 867 2191
Contact: Sam Ponzo
Description: See Spectrogon AB .

SPECTROGON AB
(Company Headquarters)
Box 2076
S-18302 Taby
Sweden
Description: Manufactures & designs interference filters for IR, visible & UV spectral regions; narrow and bandpass, long/shortwave-
pass, isolation-line, & ND filters; atmospheric
windows; diffraction gratings; AR & metallic
coatings.

SPECTROLAB INTERNATIONAL LTD.
P.O. Box 25
Newbury
Berkshire
England RG16 8BQ
United Kingdom
Voice Phone: (44)0635 248080
Fax Phone: (44) 0635248745.
Contact: Bill Vince.
Description: Manufacture holographic sys-
tems including laser tables, optics, opti and
laser positioning systems.

SPECTRON DEVELOPMENT LABORA·
TORIE
1582 Parkway Loop
Suite B
Tustin
CA92680
USA
Voice Phone: (714) 566 9060
Fax Phone: (714) 566 9055
Description: Scientific holography research;
Interferometry; Thermoplastic recording
material research; Holographic non-destruc-
tive testing.

SPINDELER & HOYER GmbH
Postfach 3353
Koenigsallee 23
Goettingen
D-34oo
Germany
Contact: Rainer Lessing
Description: Manufactures H-l's.

STALLARD, PENN
Holospace
P.O. Box 4651
Chicago
IL60680
USA
Voice Phone: (312) 942 9027
Fax Phone: (708) 866 1808
Contact: Penn Stallard
Description: Noted for combining reflection
holography with bronze sculpture Founder,
Holographic Archive of Living Presidents.
First recording of former President Ronald
Reagan is the first holograms to be accepted
by the National Portrait Gallery.

STARCKE, KY.
(Main Office)
P.O. Box 22
SF-32811
Peipohja,
Finland
Voice Phone: (358) (39) 360700
Fax Phone: +(358) (39) 367230
Contact: Mr. Ari-Veli Starcke
Description: Starcke KY is the leading com-
pany selling holograms in Scandinavia

STARLIGHT HOLOORAPHIC INC.
73 Stable Way
Kanata
Ontario K2M 1 A8
Canada
Voice Phone: (613) 235 0440
Fax Phone: (613) 592 7647
Contact: Stephen Leafloor
Description: Starlight Holographics Inc. is
one of Canada's leading representatives of
the international holographic community, in
both stock and custom holography. Its subsid-
iaries, Starmagic Holographic Gallery are
currently opening across Canada.

STEINBICHLER OvrOTECHNIK GMBH
Am Bauhof4
Neubeuern
D-8201
Germany
Voice Phone: (49) 8035 1010
Contact: Dr. H. Steinbichler
Description: Holographic analyzers,

STEPHENS, ANAIT
1685 Fernald Point Lane
Santa Barbara
CA93108
USA
Voice Phone: (805) 969 5666
Contact: Anait Stephens
Description: Artist in hands-on holography;
reflection holography, portraiture, commis-
sions.

STEREO WORLD MAGAZINE
5610 SE 71st Street
Portland
OR 97206
USA
Voice Phone: (503) 771-4440
Contact: John Dennis
Description: Magazine published bi-monthly
by its parent company, National Stereoscopic
Association. Each issue has 40 or more pages
and includes topics ranging from holography
to all aspects of stereoscopic imaging.

STEUER KG GmbH & Co.
Postfach 100327
Ernst May Strasse 7
Leinfelden-Echterdingen
D-7022
Germany
Voice Phone: (49)(0711) 753 143
Description: Manufacturer of hologaphic
embossing machines.

STOLTZAG
Tafernstrasse 15
CH-5405 Baden Dattwil
Switzerland
Voice Phone: (41)(056) 840151
Contact: Beat Ineichen
Description: Holographic non-destructive
testing.

STUDIO CREAIUS
204 Sheridan Avenue
Palo Alto
CA94306
USA
Voice Phone: (415) 960 6999
Fax Phone: (415) 960 6999
Contact: Stephan Gunning
Description: Custom-designed jewelry
pieces and holographic consulting services.

STUDIO FuR HOLOORAPHIE.
Waldfriedenweg 10
Eichenau
D-8031
Germany
Voice Phone: (49) 08141 70831
Fax Phone: (49) 08141 82268
Contact: Dr. Carlo Schmelzer
Description: Products: mastering and copy
services (rainbow/reflection), production of
mass-run embossed holograms and shims,
open stock images, art-pieces.

STUDIO WEIL-ALVARON
Ostra Tullgatan 8
S-211 28
Malmo
Sweden

Voice Phone: (46) 08141 70831
Contact: Lektor H. Herman Wei!.
Description: Hans Weil's inventions were
made in the period 1933-1937, while Gabor
invented holography in 1948, the laser was
invented 1962 and the first laser-illuminated
hologram was exposed as late as 1964.

SUPERBIN CO. LTD
3F-339
Section 2
Ho Ping E. Road
Taipei
Taiwan 10662
Taiwan
Voice Phone: (886)(02) 701 3626
Fax Phone: (886) (02) 7013531
Contact: Mr. Edward Hwang
Description: Exclusive Chinese representative
of Coherent (Argon, Krypton Laser, Dye
Laser); Continuum (ruby Laser, Nd: YAG
laser); Newport (optical components).Also
supply embossed hologram-manufacturing
equipment/material and consulting service.

SUPERIOR TECHNOLOGY IMPLEMENTA-
TION
Hjortekaersvej 99B
Lyngby
2800
Denmark
Voice Phone: (45)(45) 933 358
Fax Phone: (45)(45) 930 353
Contact: Knud Banck-Hanune
Description: Embossed holography; artistic
holography; education; HOE manufacturer;
Equipment and supplies.

SWEDE HOLOPRINT
Duvhoksgatan 6A
Malmo
21460
Sweden
Voice Phone: (46) (040) 898 21
Contact: Bjorn Wahlberg
Description: Artistic holography; Artistic
marketing consultant.

SWISS FEDERAL INSTITUTE OF
TECHN
Laboratory of Photoelasticity
Ramistrasse 101
Zurich
CH-8092
Switzerland
Voice Phone: (41) 0380 246 000
Contact: Walter Schumann
Description: Holographic non-destructive
testing; Scientific and industrial research.

SYDNEY COLLEGE OF THE ARTS
Dept. of Holography
58 Allen Street
Glebe
Sydney
Australia
Description: Holography courses.

SYNCHRONICITY HOLOGRAMS
Box 4235
RR 1
Lincolnville
ME04849
USA
Voice Phone: (207) 7633182
Contact: Arlene Jurewicz
Description: Synchronicity Holograms provides outreach educational presentations on all aspects of holography for primary grades through high school. Available for workshops and presentation on education. Research on educational aspects of holography.

T.A.1. INCORPORATED
12021 South Memorial Parkway
Huntsville
AL35803
USA
Voice Phone: (205) 8814999
Fax Phone: (205) 880 8041
Contact: Loy Shreve
Description: Manufacture and testing of optic and laser inspection equipment.

TAMA ART UMVERSITY
Department of Physics
1723 Yarimizu
Hachiou ji-shi
Tokyo
Japan
Voice Phone: 0426 76 8611
Fax Phone: 0426762935
Contact: Mr. Hidetoshi Katsuma
Description: Research on Holographic TV, Holographv Movie

TECHNICAL UNIVERSITY @ EINDHOVEN
Faculty of Architecture
Calibre Institute
P.O. Box 513
Eindhoven
NL-5600MB
Netherlands
Contact: Geert T. A. Smelzer
Description: Academic research, computer generated holograms

TECHNICAL UNIVERSITY OF BUDAPEST
Institute of Precision Mechanics
Applied Biophysics Laboratory
Budapest
H-1621
Hungary
Contact: Pal Greguss
Description: Medical holography research.

TECHNICAL UNIVERSITY OF WROCLAW.
Institute of Physics
Wybrvzeze Wyspianskiego 27
Wroclaw
PL-50-370
Poland
Contact: Henryk Kasprzak
Description: Academic, scientific research

TECHNOLAS LASER TECHNIK GmbH
Lochhamer Schlag 19
Graefelfing
8032
Germany
Voice Phone: (49) (089) 8 545040
Fax Phone: (49) (089) 854561
Description: Lasers, medical

TEXTILE GRAPHICS, INC.
(Main Office).
10884 South Street
P.O. Box 68
Nunica
MI49448
USA
Voice Phone: (616) 842 5626
Fax Phone: (616) 842 5653
Contact: Jan Bussard,
Description: Holographic stock or custom shapes for do-it-yourself permanent adhesion to texttiles and other substrates--special edges prevent delamination, withstand 100 washings/dryings. Worldwide distributors sought!

THE ART INSTITUTE OF CHICAGO
Holography Department
Columbus Drive @ Jackson Bouleva
Chicago
IL 60603
USA
Voice Phone: (312) 443-3883
Contact: Ed Wesley
Description: Offers an MFA degree with concentration in holography. Equipped with three tables, with one containing a stereogram printer for computer-generated imagery or views from life. One instructor, three graduate assistants.

THIRD DIMENSION ARTS INC.
1241 Andersen Drive
Suites C & D
San Rafael
CA94901
USA
Voice Phone: (415) 485 1730.
Fax Phone: (415) 485 0435.
Contact: Tim LaDuca.
Description: Third Dimension Arts Inc. manufacturers of dichromate jewelry, gifts, and 3-D arts trademark hologram watches. Suppliers to: the gift, jewelry, museum, and entertainment industry; (licencing) markets. Custom designs welcome!

THREE-D LIGHT GALLERY
109 The Commons
Ithaca
NY 14850
USA
Voice Phone: (607) 273 1187.
Contact: Jonathan Back
Description: Artistic holography; holography gallery.

TJING LING INDUSTRIAL RESEARCH.
130 Keelung Road
Section III
Taipei
Taiwan
Voice Phone: (86)(20704 1856
Description: Fine art originals.

TNO INSTITUTE OF APPLIED PHYSICS
Department of Optics
P.O. Box 155
NL-2600 AD Delft
Netherlands
Contact: Ruud L. van Renesse
Description: Academic and industria

TOKAI UNIVERSITY
Department of Electro Photo Opti
1117 Kitakaname Hiratsuka City
Kanagawa
259-12
Japan
Contact: Hidetoshi Katsuma.
Description: Artistic holography

TOKIMEC INC.
Hanoh Plant Electronic Lab.
257 Maehara
Yaoroshi, Hanoh
Sailwna
357
Japan
Contact: Takatsune Okada
Description: Hologram manufacturing.

TOKYO INSTITUTE OF TECHNOLOGY
Imaging Science and Engineering
4259 Nagatsuda
Midori-ku
Yokohama
227
Japan
Voice Phone: 045 -922-1111
Fax Phone: 045 9211492
Contact: Mr. Masahiro Yamasuchi
Description: Holographic Display, 3-D
Imaging Science. Also Dr. Toshio Honda.

TOPAC GmBH HOLOORAPHY
Auf'm Eickholt 47
4830 Gutersloh 1
Germany
Voice Phone: (49) 05241 580192
Fax Phone: (49) 0 52 41 58549
Contact: Dr. Werner Schumacher
Description: Embossed holography full
service. Product and marketing concepts, cre-
ation of artwork and modelling. Production
of holograms and application to specific prod-
ucts. Packaging and distribution service.

TOPACK GESELLSCHAFT FuR TON-
TRAEG
Carl-Bertelsmann-Str 161, Abt.
4830 Giitersloh 1
4830
Germany
Voice Phone: (49) 5241803302
Fax Phone: (49) 5241 79884
Description: Holograms,Holographic projects,
Nometum packings,Packagings

TOPCONInc.
75-1
Hasunuma-machi
Ttabas-i-ku
Tokyo
174
Japan
Voice Phone: 03 3966 3141
Fax Phone: 03-3966-2140
Contact: Mr. Reiji Hashimoto
Description: Hologram manufacturer.

TOPPAN PRINTING CO., LTD.
Image Technology Laboratory
1-3-3 Suido
Bunkyo-ku
Tokyo
Japan
Voice Phone: 03 3817 2873
Fax Phone: 03 56847600
Contact: Mr. Teiichi Nishioka
Description: Research in Holography, 3-D
TV, Computer Graphics

TOPPAN PRINTING CO., LTD.
Technical Research Institute
4-2-3 Takanodai-minarni
Sugito-machi Kita-Katsuskika-gun
Saitama
345
Japan
Voice Phone: 0480-34 1011
Fax Phone: 048034 1017
Contact: Mr. Fujio Iwata
Description: Holographic Display, manufac-
turer of embossed holograms, Lippman, and
Multiplex. Also Mr. Susumu Takahashi and
Mr. Toshiki Toda at this address.

TOTAL REGISTER INC.
4 Production Drive
Brookfield
CT06804
USA
Voice Phone: (203) 740 0199
Fax Phone: (203) 7400177
Contact: John Gallagher
Description: Manufacturer of registration
devices for hot-stamping attachments on
embossing equipment.

TOUCHWOOD HOLOORAPHICS
50 Sugworth Lane
Radley
Abingdon
Oxon
England OX14 2HY
United Kingdom
Voice Phone: 0865735874
Contact: Mr. George W. Clare
Description: Small experimental laboratory
investigating holographic applications for
advertising and promotions in the normal
sports trade. One-off experimental commis-
sions can be undertaken. Sorry, no long runs
or repeat work.

TOWNE LABORATORIES, INC.
P.O.Box 460-HM
One U.S. Highway 206
Somerville
NJ 08876-0460
USA
Voice Phone: (908) 722 9500
Fax Phone: (908) 722-8394
Contact: Charles Ondrejik
Description: Towne Laboratories is a producer
of fine quality precision holographic photo-
plates with or without a sub-layer of IRON-
OXIDE and precision spun striation free
photoresist in sizes to 18"x 18".

TOYAMA NATIONAL COLLEGE OF
MARIT
1-2 Ebie-Neriya
Shinminato
93302
Japan
Voice Phone: 0766 86 0511
Contact: Dr. Kenji Kinoshita
Description: Holographic Stereogram

TRANSFER PRINT FOILS INC.
Holopak Technologies
P.O.Box 518
9 Cotters Lane
East Brunswick,
NJ 08816
USA
Voice Phone: (908) 238 1800
Fax Phone: (908) 238 7936
Contact: . Rod Siberine
Description: Manufacture and development
of holographic hot-stamping foils, films and
metallized holographic paper and paper
board. In-house metallizing, embossing and
design center. We offer holographic images
as well as Trans Fraction 'patterns (gratings).

TREND
Miramarska 85
Zagreb
41000
Yugoslavia
Voice Phone: (38)(041) 51142
Contact: Dalibor Vukicevic
Description: Gallery.

TRI-ESS SCIENCES:
Student Science Service
1020 West Chestnut Street
Burbank
CA 91506-1623
USA.
Voice Phone: (818) 2476910
Description: Educational materials and laser
equipment supplier.

TRIDNENSIONALE HOLOGRAMAS
Alberto
Alcocer
38-2D
Madrid
28016
Spain
Voice Phone: (34)(481) 290 745
Contact: Daniel Weiss
Description: Artistic holography; Pulsed
portraiture.

U.K. GOLD PURCHASERS d.b.a .. Holograms Unlimited
20131 Highway 59
Unit 2216
Humble
Texas
77338
USA
Voice Phone: (512) 9935211
Fax Phone: (512) 993-0467
Contact: Marvin Uram
Description: If its holographic, we have, or want it! One stop wholesale distributor for varied products the public can afford and will buy. All items at factory-to-you prices.

U.S. HOLOORAPHICS
P.O. Box E
Logan
UT84323
USA
Voice Phone: (801) 753 5775
Fax Phone: (801) 753 5876
Contact: Dave Rayfield
Description: Marketing/manufacturing for mass-produced and custom dichromate, photopolymer and embossed holograms for retail and ad speccialty needs. Stock and custom products available.

UK OPTICAL SUPPLIES
84 Wimborne Road West
Wimborne
Dorset
England BH21 2DP
United Kingdom
Voice Phone: (44)(0202) 886 831.
Fax Phone: (44)(0202) 742 236
Contact: Ralph Cullen
Description: Supplying probably the world's largest selection of Holographic/Optical components which are: best quality; best value; Designed by experienced holographers. Plus component selection and laboratory/ studio set-up advice freely available.

ULTIMATE IMAGE
P.O. Box 715
8 Holland Street
Cronulla
NSW2230
Australia
Voice Phone: (61) (02) 523 9869
Fax Phone: (61) (02) 523 9869
Contact: Donna Kourtessis

ULTRAFINE.
16 Foster Road
Chiswick
London
England W4 4NY
United Kingdom

Fax Phone: (44)(01) 995230
Description: Holographic non-destructive testing; scientific and industrial research

UNIPHASE VETREIBS-GMBH
LeiBstr.8
Feldkirche West
815
Germany
Voice Phone: (49)08063/9036
Fax Phone: (49) 08063/7663
Contact: Werner Bleckwendt
Description: Manufacturer of lasers

UNITED TECHNOLOGIES RESEARCH CENTER
Silver Lane
East Hartford
CT06108
USA
Voice Phone: (203) 727
Fax Phone: (203) 727 7852
Contact: Dr. Karl A. Stetson.
Description: UTRC electronic holography systems display real-time fringe patterns on TV comparable to photographic holography. They also provide data output for quantitative analysis. Complete systems or retrofits are available.

UNIVERSIDADE DO PORTO
Laboratorio de Fisica
Praca Gomes Teixeira
Porto
P-4000
Portugal
Contact: Oliverio Soares
Description: Holographic non-destructive testing; Academic holography research.

UNIVERSITA DI ROMA
La Sapienza
Dipartimento di Fisica
Piazzale Aldo Moro 2
Rome
1-00185
Italy
Contact: Paolo De Santis
Description: Scientific research

UNIVERSITE DE NEUCHATEL
Institut de Microtechnique
2, rue A L Breguet
Neuchatel
CH-2000
Switzerland
Voice Phone: (41)(038) 246 000
Contact: Rene Dandliker
Description: Industrial research.

UNIVERSITY OF TSUKUBA
Institute of Art & Design
I-I, Tennodai
Tsukuba305
Japan
Voice Phone: (81) 298 53 2883
Fax Phone: (81) 298 536508
Contact: Shunsuke Mitamura
Description: Artistic holography, holography education.

UNIVERSITY OF ABERDEEN
Dept. of Engineering
Kings College
Aberdeen
Scotland AB9 2UE
United Kingdom
Description: Holographic non-destructive testing, industrial research.

UNIVERSITY OF ALABAMA
University at Huntsville
Center for Applied Optics
Huntsville
AL35899
USA
Voice Phone: (205) 895 6030.
Fax Phone: (205) 895-6618.
Contact: Chandra Vikram
Description: Scientific holography research, NDT.

UNIVERSITY OF ALICANTE
Department of Applied Physics
Centro de Holografia
Facultad de Ciencias
Alicante Apdo 99
Spain.
Voice Phone: (34)(566) 1200 ext 114
Contact: A. Fimia.
Description: Artistic holography; HOEs; workshops.

UNIVERSITY OF ARIZONA
Optical Science Center
Tucson
AZ 85721
USA
Voice Phone: (602) 621 6997
Contact: Robert Shannon
Description: Industrial and scientific Holography research; Holographic interferometry; Holographic non-destructive testing.

UNIVERSITY OF BOLOGNA
via Fiumazzo 347
Belricetto
(RA) 1-48010
Italy
Contact: Pier Luigi Capucci
Description: Artistic holography research & education.

UNIVERSITY OF CALIFORNIA
University at San Diego
Dept. Electrical & Computer Engineering
La Jolla
CA92093
USA
Voice Phone: (619) 534 2230
Contact: Sing Lee.
Description: Holography research

UNIVERSITY OF DAYTON
Research Institute
300 College Park
Dayton
OH45469
USA
Voice Phone: (513) 229 3221
Contact: Lloyd Huff
Description: Scientific research, Industrial research; courses.

UNIVERSITY OF MICHIGAN
College of Engineering,
Chrysler Center
Ann Arbor
MI 48109-2092
USA
Voice Phone: (313) 7635464
Contact: Peter M. Banks
Description: Holographic interferometry; Particle measurement; Holographic scientific & industrial research; Holographic nondestructive testing.

UNIVERSITY OF MICHIGAN
Department of Electrical Engineering
Room 1108
EECS Building
Ann Arbor
MI 48109-2122
USA
Voice Phone: (313) 764 9545
Fax Phone: (313) 763 1503
Contact: Emmet Leith
Description: Scientific holography research; Design H.O.E.s; Courses on holography.

UNIVERSITY OF MUNICH
Institute of Medical Optics
Thresienstrasse 37
Munich 2
D-8000
Germany
Contact: R. Rohler
Description: Medical holography; scientific holography research.

UNIVERSITY OF MÜNSTER
Ear, Nose and Throat Clinic
Kardinal von Galen Ring 10
Münster
D-4400

Germany
Voice Phone: (49)(0251) 836 861
Fax Phone: (49)(0251) 836 960
Contact: Gert von Bally
Description: Medical holography; Interferometry.

UNIVERSITY OF NORTH CAROLINA
Department of Cell Biology
CB 7090
108 Taylor Hall
Chapel Hill
North Carolina 27599
USA
Voice Phone: (919) 966-2941
Contact: Dr. Arnmasi Periasamy
Description: Medical holography.

UNIVERSITY OF OXFORD
Holography Group
Department of Engineering Scienc
Parks Road
Oxford
England OXI 3PJ
United Kingdom
Voice Phone: (44) 865 273 099
Fax Phone: (44) 865 273 305
Contact: Paul Hubel
Description: Scientific holography research; workshops.

UNIVERSITY OF ROCHESTER
Institute of Optics
Rochester
NY 14627
USA
Voice Phone: (716) 2755248
Contact: Dr. Duncan Moore
Description: Scientific and industrial holography research; interferometry; particle testing & measurement.

UNIVERSITY OF SOUTHERN CALIF.
Department of Physics
University Park
Los Angeles
CA 90089-0484
USA
Voice Phone: (213) 7401134
Contact: Jack Feinberg
Description: Scientific holography research; Interferometry.

UNIVERSITY OF STRATHCLYDE
Mechanical Engineering Group
Glasgow
Scotland United Kingdom
Contact: P. Waddell
Description: Scientific research; industrial research.

UNIVERSITY OF STUTTGART
Institute of Applied Optics
Pfaffenwaldring 9
Stuttgart 80
D-7000
Germany
Voice Phone: (49)(0711) 685 6075
Contact: Hans Tiziani.
Description: Scientific holography research; interferometry.

UNIVERSITY OF TOKYO
Faculty of Engineering
Hongo 7-3-1
Bunkyo-ku
Toyko
Japan
Contact: T. Uyemura
Description: Scientific and Medical holography research; Interferometry.

UNIVERSITY OF TOKYO
Institute of Industrial Science
7-22-1, Roppongi,
Minato-ku
Tokyo
106
Japan
Voice Phone: 0334026231
Fax Phone: 03 34025078
Contact: Dr. Joji Hamasaki
Description: Research

UNIVERSITY OF WISCONSIN
College of Engineering
432 North Lake Street
Madison
WI 53706
USA
Voice Phone: (608) 263 7427
Fax Phone: (608) 263 3160
Contact: Francis P. Drake
Description: Courses on Laser Systems Design. Covers laser operation, techniques for using & modifying laser output; types of lasers; with material on scanning, modulation, & detection of laser radiation; designing practical laser systems.

UNIVERSITY OF ZAGREB
Institute of Physics
Bijenicka 46
Zagreb,
41000
Yugoslavia
Voice Phone: (38)(041) 271 211
Description: Industrial, Holographic nondestructive testing; medical holography.

LWVERSITE LAVAL
Dept. Physique-COPL
Pavilion Vachon
Ste-Foy
Quebec
G1K 7P4 Canada
Contact: Roger A. Lessard
Description: Holography education; workshops.

UNIVERSITIlT ERLANGEN - NURNBERG
Physikalisches Institute
Erwin-Rommel Strasse 1
0-8520 Erlangen
0 -8520
Germany
Voice Phone: (49)(09131) 857 408
Contact: Adolf Lohmann
Description: Scientific holography research;
HOE; computer generated holography.

UNIVERSITe DE FRANCHE-COMTE
Laboratoire d 'Optique
L.A. 214
UFR Sciences et des Techniques
F-25030 Besancon Cedex
France
Description: Scientific holography research.

UNIVERSITe DE PARIS-SUD
Institute d'Optique
Orsay
F-91405
France
Voice Phone: (33)(9) 416750
Contact: Serge Lowenthal
Description: Scientific holography

UNIVERSITe LOUIS PASTEUR
Photonics Systems Laboratory
7 rue de L'Universite
Strasbourg
67000
France
Voice Phone: (33) (88) 355750.
Contact: ; P. Meyrueis
Description: Scientific holography research.

UNTERSEHER & ASSOCIATES
HOLOORAPHY
709 1/2 West Glen Oaks Blvd.
Glendale
CA91202
USA
Voice Phone: (818) 549 0534
Contact: Fred Unterseher.
Description: Artistic holography and holography education.

VAN LEER METALLaED PRODUCTS
(USA)
P.O. Box 9321
119 Herbert Street
Framingham
MA01701
USA
Voice Phone: (508) 820 3406
Fax Phone: (508) 875 0057
Contact: Randy Jacobs
Description: Manufacturer of HoloPRISM'
holographic metallized papers. Van Leer
produces diffraction patterns or multi-dimensional holographic images on direct metallized papers. Existing patterns or customized
images are available.

VINCENNES UNIVERSITY
1002 North First Street
Vmcennes
IN 47591
USA
Voice Phone: (812) 885-5294
Contact: Richard Duesterberg
Description: Offering holography workshops
for high school teachers, & college level
courses in holography. We have 4 research-
grade optical tables, as well as argon and
krypton lasers. Call for more details

VINTEN ELECTRO OPTICS LTD
Unit 28
Ashfield Way
Whetstone
Leicester
Eng'and LE8 3NU
United Kingdom
Voice Phone: (44)(533) 867 110
Description: ManufacturelDesign mirrors
and optics.

VISIONS OF THE FUTURE
1040 N. Kings Highway
Suite 717
Cherry Hill
NJ08034
USA
Voice Phone: (609) 667-5622
Fax Phone: (609) 667-6289
Contact: Alan Schorr
Description: Distributors for all of the holographic merchandise that you will ever need.
Prompt, knowledgeable servis our specialty.
Direct from factory prices. There is no need
to call anyone else.

VNIKTK KULTURA
ul.lntusiastov 34
Moskow
105118
Russia
Contact: O.B. Serov
Description: Marketing consultant.

VOLKSWAGEN AG
Forschung und Entwicklung
Messtechnik D-3180
Wolfsburg 1
Germany
Voice Phone: (49)(05) 361 221
Contact: Armin Felske
Description: Industrial research, Interferom-
etry; Holographic non-destructive testing

VOLVO-FLYGMOTOR
S-461
81 Trollhattan
Sweden
Voice Phone: (46) (0520) 94471
Contact: Contt: Robert Frankmark.
Description: Holographic non-destructive
testing.

WASEDA UNIVERSITY
Dept. of Applied Physics
School of Science and Engineerin
Tokyo
160
Japan.
Voice Phone: (81)(03) 209 321
Description: Medical holography research.

WAVE MECHANICS
1535 North Ashland Avenue
Second Floor
Chicago
IL60622
USA
Voice Phone: (312) 829 WAVE
Contact: Deni Drinkwater-Welch
Description: Artistic holographer; silver ha-
lide transmission and reflection; consultant.

WAVEFRONT TECHNOLOGIES
7343 Adams Street
Paramount
CA90723
USA
Voice Phone: (310) 634 0434
Contact: Joel Petersen
Description: Holographic fine artist

WAVEFRONTS
1223 7th Avenue
San Francisco
CA 94122
USA
Voice Phone: (415) 664 0694
Contact: Louis Brill
Description: Involved in developing & expanding market & sales efforts for holographic
retaiVwholesale product lines. Assist in
preparation of promotions and collateral sales
materials, identify potential sales markets &
implementation of sales.

WHILEY FOILS LIMITED
Firth Road
Houston Industrial Estate
Livingston
st Lothian
Scotland EH54 5DJ
United Kingdom.
Voice Phone: (44) (0506) 38611
Fax Phone: (0506) 38262.
Contact: B J itch
Description: Whiley Foils Limited is a lon-
gestablished manufacturer of stamping foils.
We have developed special base materials for
Holographic embossing and market these and
Holographic foils worldwide.

WHITE LIGHT WORKS, INC.
P.O. Box 851
Woodland Hills
CA91365
USA
Voice Phone: (818) 7031111
Fax Phone: (818) 7031182
Contact: Jerry Fox.
Description: Full-service holographic pro-
duction company specializing in low cost
embossed holograms and Multiplex "People
Stopper" holographic displays for trade shows
and POP applications. Three retail outlets in
Southern California.

WHITE TIGER HOLOORAMS
Johannes Verhulststraat 45
1071 MS Amsterdam
Netherlands
Voice Phone: (31)(20) 797 270
Fax Phone: (31)(20) 790 896
Contact: Neil Walker
Description: Partners' experience 30 years in
advertising, and 12 in holography. Established 1986. White Tiger Holograms has
completed major projects in embossed, dichromate and silver halide, as well as security
and production consultancy.

WHOLE HOGRAPHY
4142 Bellefontaine Street
Houston
TX 77025-1105
USA
Voice Phone: (713) 667 3325
Contact: Michael E. Crawford
Description: Artistic holography

THE WHOLE PICTURE
A Gallery of Holography
634 Parkway
Gatlinburg
TN
37738 USA
Voice Phone: (615) 4363650
Fax Phone: (615) 4368974
Contact: Jim Kelly
Description: Full line of film and plate holo-
grams. Full line of holographic jewelry items.

Holographic Head-Up Display

Holographic Night Vision Goggle

Holographic Helmet-Mounted Display

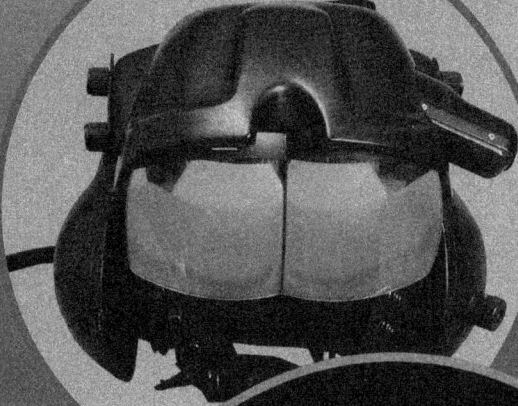

Hughes Aircraft Company is applying 20 years of experience to commercial holographic products

Holographic Automotive Stop Lamp

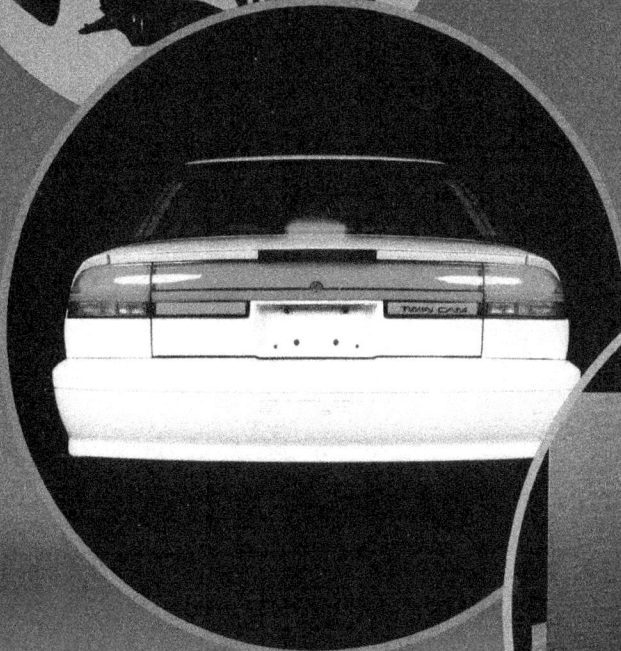

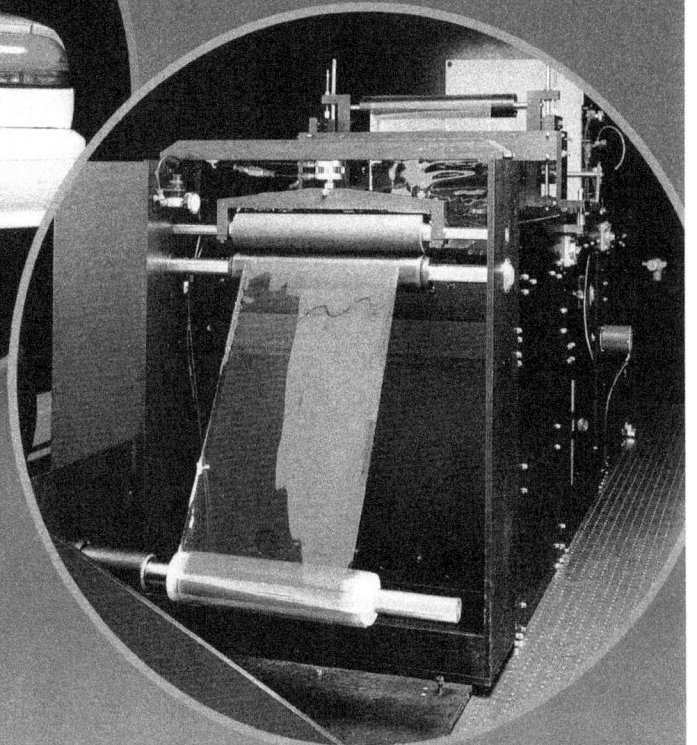

High-Volume Production Using DuPont OmniDex® Photopolymer

Starting in 1972 with pioneering work at its Malibu Research Laboratories, Hughes Aircraft Company has continuously pushed the state of the art to develop holographic optics for use in aerospace systems. A team with 200 years of total holographic experience, 75 U.S. patents, and a 4000-square-foot holographic clean-room facility is available to design, develop, and produce holograms for your commercial applications.

Holographic Optical Elements (HOEs)

- Mirrors and reflectors
- Lenses
- Lens arrays
- Diffusing screens
- Filters
- Gratings
- Complete optical design, development, and production capabilities

Commercial Applications

- Security
- Decorative
- Advertising
- Publications
- Mastering capabilities for 2-D, 3-D, and multiplex holograms

High-Rate Production with DuPont OmniDex® Photopolymer

- Replication from flat master hologram
- Image holograms or HOEs
- Transmission or reflection copies
- Production starting third quarter 1993

OmniDex® is a registered trademark of E.I. duPont de Nemours and Company

To discuss your holography application or for additional information on production services, please contact John Gunther at (310) 334-7753 or by facsimile at (310) 334-7756.
PO Box 92426 RE / R1 / Mail Station V500
Los Angeles, CA 90009

HUGHES

HUGHES POWER PRODUCTS

WORCESTER POLYTECHNIC INSTITUTE
Mechanical Engineering Departmen
100 Institute Road
Worcester
MA01609-2280
USA
Voice Phone: (508) 8315536
Fax Phone: (508) 831 5483
Contact: Ryszard Pryputniewicz
Description: Scientific, Medical & Industrial
holography research; Interferometry; Holographic non-destructive testing.

WORLD ART PROJECT
10247 40th Street West
Webster
MN55088
USA
Voice Phone: (507) 744 2913
Description: Scientific and holography
research; holographic non-destructive testing;
interferometry.

WRIGHT PATTERSON AIR FORCE
BASE.
WL/FIBG
Dayton
OH45433
USA
Voice Phone: (513) 255-5159
Fax Phone: (513) 2561442
Contact: Gene Maddux
Description: Scientific and holography
resear

WYKOCorp.
2650 East Elvira Road
Tucson
AZ 85706-7123
USA
Voice Phone: (602) 741-1044
Fax Phone: (602) 294-1799
Contact: James Wyant
Description: Scientific holography research;
Interferometry and analysis.

X-IAL
Les Algorithmes"
Parc d'lnnovation
IIIkirch
F-67400
France
Voice Phone: (33) 88 674490
Fax Phone: (33) 88 674066
Contact: Dr. Christian D. Liegeois
Description: Stereograms for embossing;
HOEs designed and manufactured

In Collaboration With Artists
To Produce
World Class Art Holography
Setting The Standard In
Full Color

H O L O G R A P H I C
I   M   A   G   E   S

"Limited Editions"
For Color Brochure:
Holographic Images Inc.
1301 Dade Blvd. · Miami Beach, Fl · 33139

Ph: (305) 531-5465 · Fax: (305) 531-3029

14

Businesses by Category

In this chapter we list the businesses by their primary business activities. You can find a business quickly here and then go to the main listing (in the chapter before this one) to get a full description of their services and products.

Mastering and Single copy and holograms:

3D VISION (Germany)
3DI (United Kingdom)
A.H. PRISMATIC, INC. (USA)
A.H. PRISMATIC, LTD. (United Kingdom)
ACME HOLOGRAPHY (USA)
AD 2000, Inc. (USA)
ADVANCED HOLOGRAPHIC LABS (United Kingdom)
ADVANCED HOLOGRAPHIC LABS (USA)
ADVANCED IMAGING TECHNOLOGIES (United Kingdom)
AKS HOLOGRAPHIE-GALERIE GmbH. (Germany)
AMERICAN BANK NOTE HOLOGRAPHICS, (USA)
ANOTHER DIMENSION. (USA)
AP HOLOGRAFIKA STUDIO (Hungary)
APPLIED HOLOGRAPHICS, PLC. (United Kingdom)
ARBEITSKREIS HOLOGRAFIE B.V. (Germany)
ARCHEOZOIC INCORPORATED. (USA)
ARCHITECfURAL GLASS & HOLOGRAPH (United Kingdom)
ARMIN KLIX HOLOGRAPHIE --Germany
ART & DESIGN GRAPHICS (Russia, RF)
ART FREUND HOLOGRAPHY (USA)

ART INSTITUTE OF CHICAGO (USA)
ARTBRIDGE MANAGEMENT (Germany)
ARTIGLIOGRAPHY CO. (USA)
ARTKITEK (USA)
ASAHI GLASS CO. (Japan)
ASCOT LASER PICTURE STUDIO (United Kingdom)
ATELIER HOLOGRAPHIQUE DE PARIS (France)
AUSTRALIAN HOLOGRAPHICS PTY LTD. (Australia)
BARR & STROUD LTD (United Kingdom)
BENYON, MARGARET(United Kingdom)
BERKHOUT, RUDIE (USA)
BOB MADER PHOTOGRAPHY. (USA)
BOOTH,ROBERT A(USA)
BOYD, PATRICK (United Kingdom)
BRIDGESTONE GRAPHIC (USA)
BRIGHTON IMAGECRAFT. (United Kingdom)
BURNS HOLOGRAPHICS LTD (USA)
BYRNE,KENNETH(USA)
CAMBRIDGE STEREOGRAPHICS GROUP. (USA)
CASDIN-SILVER HOLOGRAPHY. (USA)
CFC/APPLIED HOLOGRAPHICS (USA)
CHERRYOVITCAL HOLOGRAPHY. (USA)
CHROMAGEM INC. (USA)
COBURN CORPORATION (USA)
COSSETIE, MARIE ANDREE (Canada)

CROSS, LLOYD (USA)
DAVID SCHMIDT HOLOGRAPHY USA.
DEEM,REBECCA(USA)
DEEP SPACE HOLOGRAPHICS (Canada)
DIALECfICA AB (Sweden.)
DIAURES S.A. HOLOGRAPHIC DIVISIO (Italy)
DUSTON HOLOGRAPHIC SERVICES (Canada)
DUTCH HOLOGRAPHIC LABORATORY. (Netherlands)
ElF PRODUCfIONS (France.)
ED WESLY HOLOGRAPHY. (USA)
ELECfRO OVITC CONSULTING SERVICE (USA)
EMPAQUES Y EVOLTURAS (Mexico) FLATIRON STUDIO (USA)
FLEXCON COMPANY, INC. (USA)
FOREIGN DIMENSION (Hong Kong)
FORNARI, DAVID (USA)
FRINGE RESEARCH HOLOGRAPHICS (Canada.)
FTI JOFFE (Russia)
GENERALFEROE(USA)
(GENERAL HOLOGRAPHICS, INC. (Canada)
GERALD MARKS STUDIO. (USA)
GRAY SCALE STUDIOS LTD. (USA)
HARRIS, NICK (USA)
HELLENIC INSTITUTE OF HOLOGRAPHY (Greece)

Mastering & Single Copies (cont)

HIGH TECH NETWORK (Sweden)
HM-Holographie (Germany)
HOECHST CELANESE CORPORATION (USA)
HOLAGE (USA.)
HOLICON CORPORATION. (USA)
HOLO GMBH HOLOORAFIELABOR (Germany)
HOLO-DIMENSIONS INC (Canada)
HOLO-IMAGES, INC. (USA)
HOLO-LASER. (France)
HOLO-SERVICE. (Switzerland)
HOLO-SERVICE.FRIES (Switzerland)
HOLO-SOURCE CORPORATION (USA)
HOLO-SPECTRA (USA)
HOLOCOM HOLOGRAPHIE (France.)
HOLOCRAFT INTERNATIONAL (USA)
HOLOCRAFTS EUROPE LIMITED. (United Kingdom)
HOLOCRAFTS: CANADIAN (Canada)
HOLOFAR LAB (SRL) (Italy)
HOLOGRAMM WERKSTATT & GALE- RIE (Switzerland)
HOLOORAMS 3D (United Kingdom)
HOLOGRAMS FANTASTIC & ILLU- SIONS (Australia)
HOLOGRAPHIC DIMENSIONS, INC. (USA)
HOLOGRAPHIC IMAGES INC. (USA)
HOLOGRAPHIC INDUSTRIES (USA)
HOLOGRAPHIC RESEARCH PTY LTD. (Australia)
HOLOGRAPHIC STUDIO, THE (Canada)
HOLOGRAPHIC STUDIOS (USA)
HOLOGRAPHICS (UK) LTD. (United Kingdom) HOLOGRAPHICS NORTH INC. (USA)
HOLOGRAPHIE LABOR (Germany)
HOLOGRAPHY CENTER OF AUSTRIA.. (Austria)
HOLOGRAPHY INSTITUTE (USA)
HOLOMART (United Kingdom)
HOLOMAT (USA)
HOLOMORPH VISUALS, INC. (Canada)
HOLOPRESS KG (Germany)
HOLOPRINT ROSOWSKI (Germany)
HOLOPRODUCTION. (France)

HOLOTEC (Germany)
HOLOTEC BIRENHEIDE (Germany)
HOLOVISION AB. (Sweden)
HUGHES AIRCRAFT CO.- (USA)
IMAGEN HOLOORAPHY, INC (USA)
ISHII, MS. SETSUKO (Japan)
JAYCO HOLO. (United Kingdom)
K.C. BROWN HOLO. (United Kingdom)
KAC, EDUARDO (USA)
KAUFFMAN, JOHN (USA)
L.A.S.E.R. CO. (USA)
LABOR FuR HOLOORAFIE (Germany)
LAMINEX/HIGH TECH UK LTD. (United Kingdom) LASART LTD. (USA)
LASER LIGHT DESIGNS (USA)
LASER LIGHT EXPRESSIONS PTy. LTD (Australia)
LASER LIGHT IMAGE. (United Kingdom)
LASER LIGHT LTD. (USA)
LASERFILM ECKHARD KNUTH (Germany)
LASERION HANDELS GMBH (Germany)
LASERSMITH, INC. (USA.)
LASERWORKS (USA) LASIRIS INc.(Canada)
LAZA HOLOGRAMS (United Kingdom)
LES PRODUCTIONS HOLOLAB(Canada)
LEVINE, CHRIS (United Kingdom)
LIBERATO, PAUL (USA)
LIGHT FANTASTIC PLC-United Kingdom
LIGHT IMPRESSIONS, INC. (USA.)
LUND INSTITUTE OF TECH. (Sweden)
MACSHANE HOLOORAPHY (USA.)
MAN ENVIRONMENT, INC. (USA.)
MARTINSSON ELEKTRONIK (Sweden)
MC CORMACK, SHARON (USA)
MGD PRODUCTIONS (Canada)
MILLER, PETER (USA)
MULTI-PURPOSE HOLOGRAMS--France
MUNDA Y SPATIAL IMAGING. (United Kingdom)
MU's LASER WORKS (Canada V8Z)
NAIMARK, MICHAEL (USA)
NATIONAL HOLO. STUDIOS (USA)
NEOVISION PRODUCTIONS. (USA.)
NEW YORK HOLO. LABS (USA)
NORTH AMERICAN HOLO. (USA.)
NORTH DANCER HOLOGRAPHIC-- USA
NORTHERN LIGHTWORKS (USA)
OlE RESEARCH (USA)

ODHNER HOLOGRAPHICS (USA)
OP-GRAPHICS   (HOLOORAPHY) LTD. (United Kingdom.)
ORIEL CORPORATION (USA)
OXFORD HOLOORAPHICS. (United Kingdom)
PACIFIC HOLOGRAPHICS, INC. (USA)
PARKER, JULIE WALKER (USA)
PHOENIX HOLOGRAMS (Germany)
PHOTON LEAGUE OF HOLOGRAPHERS. (Canada)
PINK, PATTY (USA)
POINT OF VIEW DIMENSIONS, LTD. (USA)
POINT SOURCE PRODUCTIONS (USA)
POLYMER HOLOORAPHICS, INC (USA)
POLYMER IMAGE (USA)
PORTSON, INC. (LASER IMAGES) USA.
RANDAZZO, DEAN (USA)
RANDY JAMES HOLOGRAPHY (USA)
RAVEN HOLO. LTD. (United Kingdom)
RED BEAM, INC (USA)
RICHARD BRUCK HOLO (USA)
RICHMOND HOLOGRAPHIC STUDIOS (United Kingdom)
ROBERT SHERWOOD HOLO DESIGN (USA)
ROY, JEAN (Canada)
ROYAL COLLEGE OF ART. (United Kingdom)
SOUTHERN INDIANA HOLOGRAPHICS (USA)
SPECTRATEK CORPORATION (USA)
SPINDELER & HOYER GmbH (Germany)
STALLARD, PENN (USA)
STEPHENS, ANAIT (USA)
STUDIO FuR HOLOGRAPHIE. Germany
SUPERIOR TECHNOLOGY IMPLEMEN- TATI (Denmark)
SWEDE HOLOPRINT (Sweden)
THREE-D LIGHT GALLERY (USA)
TOUCHWooD HOLOORAPHICS (United Kingdom)
U.S. HOLOGRAPHICS (USA)
UNTERSEHER & ASSOCIATES (USA)
WAVE MECHANICS (USA)
WAVEFRONT TECHNOLOGIES (USA)
WHITE LIGHT WORKS, INC. (USA)
WHITE TIGER HOLO. (Netherlands)
WHOLE HOGRAPHY (USA)
X-IAL (France)

Businesses That Produce Holograms In Large Quantity:

AD 2000 (USA)
ADVANCED HOLOORAPHIC (USA) AKS HOLOORAPHIE-GALERIE GmbH. (Germany)
AMERICAN BANK NOTE HOLOGRAPHICS (USA)
APPLIED HOLOGRAPHICS, PLC. (United Kingdom)
ARCHITECfURAL GLASS & GRAPH (United Kingdom)
AUSTRALIAN HOLOGRAPHICS PTY LTD. (Australia)
BEDDIS KENLEY (MACHINERY) LTD. (United Kingdom)
BRIDGESTONE GRAPHIC (USA)
BRIGHTON IMAGECRAFf. (United Kingdom)
BURNS HOLOGRAPHICS LTD (USA)
BYRNE,KENNETH(USA)
CFC/APPLIED HOLOGRAPHICS (USA)
CHERRYovnCAL HOLOGRAPHY (USA.)
CHROMAGEM INC. (USA)
COBURN CORPORATION (USA)
CREATIVE LABEL (USA)
DAI NIPPON PRINTING CO LTD. (Japan)
DAVID SCHMIDT HOLOGRAPHY -USA.
DE LA RUE HOLOGRAPHICS LTD.
DIFFRACfION COMPANY (USA)
DUTCH HOLOORAPHIC LAB (Netherlands)
EMPAQUES Y EVOLTURAS (Mexico)
FANTASMA INC (USA)
FLEXCON COMPANY, INC. (USA) FOREIGN DIMENSION (Hong Kong)
FUlITSU LABORATORIES LTD. (Japan)
GARDENER PROMO. MARKETING (USA)
HOECHST CELANESE CORP (USA) HOLICON CORPORATION. (USA)
HOLO GMBH HOLOORAFIELABOR (Gennany)
HOLO IMPRESSIONS INC (Taiwan)
HOLO-SOURCE CORPORATION (USA) HOLOCOR I.B.F. PRINTING (Canada) HOLOCRAFf INTERNATIONAL (USA)
HOLOCRAFTS EUROPELIMITED. (United Kingdom)
HOLOCRAFfS: CANADIAN (Canada)
HOLOORAMS FANTASTIC & ILLUSIONS (Australia)
HOLOORAPHIC DIMENSIONS(USA)
HOLOORAPHIC INDUSTRIES, INC USA
HOLOORAPHIC LABEL CONVERTING (USA)
HOLOORAPHIC RESEARCH PTY LTD. (Australia.)
HOLOGRAPHIC STUDIO, THE (Canada)
HOLOORAPHIC STUDIOS (USA)

HOLOORAPHICS (UK) LTD. (United Kingdom)
HOLOGRAPHICS NORTH INC. (USA)
HOLOGRAPHY CENTER OF AUSTRIA.. (Austria.)
HOLOMART (United Kingdom)
HOLOMORPH VISUALS, INC. (Canada)
HOLOPRESS KG (Germany)
HOLOPRODUCfION. (France)
HOLOVISION AB. (Sweden)
HUGHES AlRCRAFf CO.- (USA)
IMAGEN HOLOGRAPHY, INC (USA)
JAYCO HOLO. (United Kingdom)
K.C. BROWN HOLO. (United Kingdom)
LASART LTD. (USA)
LASER LIGHT EXPRESSIONS -Australia
LASERSMITH, INC. (USA.)
LAZA HOLOGRAMS (United Kingdom)
LETIERHEAD PRESS INC. (USA.)
LIGHT FANTASTIC PLC -United Kingdom
LIGHT IMPRESSIONS EUROPEPLC. (United Kingdom)
LIGHT IMPRESSIONS, INC. (USA.)
MARKEM SYSTEMS LTD. (United Kingdom)
MARKETING & PROMOTION CfR-Japan
MARTINSSON ELEKTRONIK . (Sweden)
MC CORMACK, SHARON (USA)
METAMORFOSI OLOGRAFIA ITALIA (Italy)
NIPPONDENSO Co., Ltd. (Japan)
NORTH AMERICAN HOLOORAPHICS (USA.)
ODHNER HOLOORAPHICS (USA) OP-GRAPHICS (HOLOGRAPHY) LTD. (United Kingdom.)
OPTICAL IMAGES (USA.)
OXFORD HOLOORAPHICS. Kingdom)
PACIFIC HOLOGRAPHICS, INC. (USA)
POINT OF VIEW DIMENSIONS (USA)
POLYMER HOLOORAPHICS, INC. (USA)
POLYMER IMAGE (USA)
PORTSON, INC. (LASER IMAGES) USA.
RAINBOW SYMPHONY INC. (USA)
REYNOLDS METALS CO. (USA)
RICHMOND HOLOORAPHIC STUDIOS (United Kingdom)
SILLCOCKS PLASTICS INT . (USA)
SPECfRATEK CORPORATION (USA)
STARLIGHT HOLOORAPHIC (Canada)
STUDIO FUR HOLOORAPHIE (Germany)
THIRD DIMENSION ARTS INC. (USA)
TOKIMEC INC. (Japan)
TOPAC GmBH HOLOORAPHY Gennany
TOPCON Inc. (Japan)
TOPPAN PRINTING CO., LTD. (Japan)
TRIDIMENSIONALE HOLO. (Spain)
U.S. HOLOORAPHICS (USA)
VAN LEER METALLIZED PRODUCTS (USA)
WHITE LIGHT WORKS, INC. (USA)
WHITE TIGER HOLO. (Netherlands)
X-IAL(France)

Business That Market Holograms (Retail And Wholesale):

3D VISION (Gennany)
A.H. PRISMATIC, INC. (USA)
A.H. PRISMA TIC, LTD. (United Kingdom)
ABRAMZIK, CURT (Gennany)
ACME HOLOORAPHY (USA)
AD 2000, Inc. (USA)
AMAZING IMAGES (USA)
AMAZING WORLD OF HOLOGRAMS. (United Kingdom)
AMERICAN BANK NOTE HOLOGRAPHICS, (USA)
AP HOLOORAFIKA STUDIO (Hungary)
APPLIED HOLOGRAPHICS, PLC. (United Kingdom)
ARMIN KLIX HOLOORAPHIE Germany
ART & DESIGN GRAPHICS (Russia, RF)
ART INSTITUTE OF CHICAGO (USA)
ARTBRIDGE MANAGEMENT (Germany)
ARTIGLIOORAPHY CO. (USA)
ATELIER HOLOORAPHIQUE DE PARIS (France)
AUSTRALIAN HOLOGRAPHICS PTY LTD. (Australia)
BRAINET CORPORATION (Japan)
BRIDGESTONE GRAPHIC (USA)
BURNS HOLOORAPHICS LTD (USA)
CHERRY OPTICAL HOLOGRAPHY (USA.)
CREATIVE HOLOGRAPHY INDEX (Germany)
DAI NIPPON PRINTING CO., LTD(Japan)
DEC-ART INC (Canada)
DEEP SPACE HOLOORAPHICS (Canada)
DIE DRITIE DIMENSION. (Germany)
DIFFRACfION COMPANY, THE (USA)
DIMENSIONS (Pakistan)
DIRECf HOLOORAPHICS. (USA)
DREAM IMAGES. (Gennany)
EDMUND SCIENTIFIC (USA.)
ELUSIVE IMAGE. (USA)
EMPAQUES Y EVOLTURAS (Mexico,)
FANTASM A INC (USA)
FISHER SCIENTIFIC. (USA.)
FLEXCON COMPANY, INC. (USA)
FOREIGN DIMENSION (Hong Kong.)
GARDENER PROMOTION MARKETING (USA)
GENERAL HOLOGRAPHICS (Canada.)
HELlOS HOLOORAPHY INC. (USA)
HELLENIC INSTITUTE OF HOLOGRAPHY (Greece)
HOLO-IMAGES (USA)
HOLO-LASER. (France)
HOLOCRAFT INTERNATIONAL (USA)
HOLOCRAFTS EUROPE LIMITED. (United Kingdom)
HOLOGRAM LAND (USA)
HOLOGRAMM WERKSTATT & GALE- RIE (Switzerland)
HOLOORAMS INTERNATIONAL (USA)

HOLOGRAPIDC MARKETING (USA)
HOLOGRAPIDC CONCEPTS, INC. USA
HOLOGRAPIDC INDUSTRIES, INC USA
HOLOGRAPIDC RESEARCH PTY LTD.
(Australia)
HOLOGRAPIDC SERVICE (Italy.)
HOLOGRAPIDCS NORTH INC. (USA)
HOLOGRAPHY CENTER OF AUSTRIA ..
(Austria)
HOLOLASER GALLERY (U.A.E.)
HOLOMART (United Kingdom)
HOLOMEDIA AB/HOLOGRAM
MUSEUM. (Sweden)
HOLOMEDIA FRANCE (France)
HOLOMEDIA INC. (Japan)
HOLOPRINT ROSOWSKI (Germany)
HOLOPRODUCTION. (France)
INTEGRAF. (USA)
JAPAN COMMUNICATION ARTS CO.
(Japan)
LAS ART LTD. (USA)
LAZA HOLOGRAMS (United Kingdom)
LAZART HOLOGRAPIDCS. (Australia.)
LAZERvnzARDRY(USA)
UGHT FANTASTIC PLC United Kingdom
UGHT IMPRESSIONS EUROPE PLC.
(United Kingdom)
UGHT IMPRESSIONS, INC. (USA.)
UGHT WAVE GALLERY (USA.)
LONE STAR ILLUSIONS (USA)
MAGIC LASER (France)
METAMORFOSI OLOGRAFIA ITAUA
(Italy)
NEOVISION PRODUCTIONS (USA.)
NEW CLEAR IMPORTS LTD. (United
Kingdom.)
NEW DIMENSION HOLOGRAPIDCS.
(Australia)
OP-GRAPIDCS LTD. (United Kingdom.)
OXFORD HOLOGRAPIDCS (United King-
dom)
PALCO'S HOLOGRAM WORLD (USA)
PHANTASTICA (Germany)
PLANET 3-D (USA)
RICHMOND HOLOGRAPIDC STUDIOS
(United Kingdom)
SCHAFFNER GALLERY (USA)
SCIENCE KIT & BOREAL LABS (USA)
SOVISKUSSTVO v/o MEZHDUNAROD-
NAYA (Russia)
STARCKE, KY. (Finland)
STARUGHT HOLOGRAPIDC (Canada)
TIDRD DIMENSION ARTS INC. (USA)
TOPAC GmBH HOLOGRAPHY (Germany
U.K. GOLD PURCHASERS (USA)
U.S. HOLOGRAPIDCS (USA)
VISIONS OF THE FUTURE (USA)
WIDTE UGHT WORKS, INC. (USA)
WIDTE TIGER HOLOGRAMS (Netherlands)
WHOLE PlcruRE (USA)

Businesses Providing Embossing Equipment & Supplies:

BEDDIS KENLEY (MACHINERY) LTD.
(United Kingdom)
BOBST GROUP (USA)
BRANDTJEN & KLUGE, INC, (USA)
BURNS HOLOGRAPIDCS LTD (USA)
CHROMAGEM INC. (USA)
COBURN CORPORATION (USA)
CROWN ROLL LEAF, INC., (USA.)
DAZZLE EQUIPMENT CO. (USA)
DUTCH HOLOGRAPIDC LABORATORY.
(Netherlands)
E.C. SCHULTZ & COMPANY, (USA)
GLOBAL IMAGES, INC. (USA)
HOLO-LASER. (France)
JAMES RIVER PRODUCTS (USA)
LEONARD KURZ GMBH (Germany)
LOUGHBOROUGH UNIV. OF TECH.
(United Kingdom)
METAPLAST ELECTROCHEMICALS
CORP. (USA)
SIDPLEY CHEMICAL CO. (USA)
STEUER KG GmbH & Co. (Germany)
TOTAL REGISTER INC. (USA)
TOWNE LABORATORIES, INC. (USA)
TRANSFER PRINT FOILS INC. (USA)
WHILEY FOILS LIMITED (United Kingdom)

Businesses Providing Lasers:

ADLAS G.m.bR & co Germany
ADVANCE PHOTONICS (India.)
AEROTECH INC. (USA)
AG ELECTRO-OPTICS LTD. (United
Kingdom)
AIMS OPTRONICS SA/NV, (Belgium)
AMERICAN LASER CORPORATION.
(USA)
BBT INSTRUMENTER APS. (Denmark)
BURLEIGH INSTRUMENTS, INC. (USA)
CAMBRIDGE LASERS, INC. (USA)
COHERENT, INC. (USA)
CORRY LASER TECHNOLOGY INC
(USA)
CVI LASER CORPORATION (USA)
DB ELECTRONIC INSTRUMENTS S.R.L.
(Italy.)
EAUNG SCIENTIFIC LTD (Canada)
ELECTECH DISTRIBUTION SYSMS
(Singapore)
ELEKTRO-PHYSIK AACHEN GmbH
(Germany)

EXCITEK INC. (USA)
FOSTEC GmbH FEINMECHANIK
(Germany)
FUn ELECTRIC CO. LTD (Japan.)
GRESSER, E., KG (Germany)
INSTITUTE OF ELECTRONICS BSSR
(Russia)
ION LASER TECHNOLOGY INC. (USA)
JODON INC. (USA)
LASER ELECTRONICS PTY. (Australia.)
LASER IONICS INC. (USA)
LASER PHOTONICS, INC. (USA)
LASER RESALE INC. (USA.)
LASER TECH INDUSTRIES (USA)
LASERMETRICS, INC. (USA)
LASING S.A., (Spain)
LENOX LASER. (USA)
LEXEL LASER, INC. (USA)
UCONIX (USA.)
UNE LITE LASER CORP (USA)
LPT LASER PHYSIKTECHNIK GmbH
(Germany)
LUMONICS INC. (Canada.)
LUMONICS LTD (United Kingdom.)
MARUBUN CORPORATION (USA)
MBB-Industrieerzeugnisse (Germany)
MELLES GRIOT (USA)
MELLES GRIOT GmbH (Germany)
MEREDITH INSTRUMENTS. (USA)
METRO LOGIC GMBH (Germany)
METROLOGIC INSTRUMENTS (USA)
MIDWEST LASER PRODUCTS (USA)
MOELLER WEDEL OPTISCHE WERK
(Germany)
MWK INDUSTRIES (USA)
NEWPORT (ASIAN OFFICE) (Japan.)
NEWPORT CORPORATION (USA.)
NEWPORT GMBH (Germany)
NEWPORT GmbH (Germany)
OMNICHROME (USA)
PHASE-R CORPORATION (USA)
ROFIN-SINAR LASER GmbH (Germany)
SIEMENS (USA) (USA)
SIEMENS LTD. (United Kingdom)
SPECTRA-PHYSICS INC. (USA)
SUPERBIN CO. LTD (Taiwan)
T.A.1. INCORPORATED (USA)
TECHNOLAS LASER TECHNIK GmbH
(Germany)
UNIPHASE VETREIBS-GMBH(Germany)

Businesses providing lab equipment
(film, tables, etc.):
ADVANCED OPTICS, INC. (USA)
AEROTECH INC. (USA)
AG ELECfRO-OPTICS LTD. (United Kingdom)
AGFA(USA)
ALPHA PHOTO PRODUCTS, INC (USA)
BRIGHTON IMAGECRAFT. (United Kingdom)
BURLEIGH INSTRUMENTS, INC. (USA)
C.ITOH & COMPANY (Japan)
CITY CHEMICAL (USA.)
EASTMAN KODAK COMPANY. (USA)
HOLOMEX LTD. (United Kingdom)
INTEGRAF. (USA)
KEYSTONE SCIENTIFIC CO. (USA)
KINETIC SYSTEMS (USA)
LASER TECHNOLOGY, INC. (USA)
MAN ENVIRONMENT, INC. (USA.)
METROLOGIC INSTRUMENTS (USA)
MICRAUDEL. (France)
MITUTOYO MEASURING INSTRUMENTS. (USA)
NEWPORT CORPORATION (USA.)
NIPPON POLAROID K.K. (Japan)
NORLAND PRODUCTS, INC. (USA)
PHOTOGRAPHERS FORMULARY(USA)
PILKINGTON OPTRONICS (United Kingdom)
POLAROID CORPORATION (USA)
SCIENCE & MECHANICS INSTRUMENTS (USA)
SHIPLEY CHEMICAL CO. (USA)
SILLCOCKS PLASTICS INTERNATIONAL (USA)
SPECTROLAB INTERNATIONAL LTD. (United Kingdom)
TOWNE LABORATORIES, INC. (USA)
Businesses Providing Optics &
HOEs:
3M--OPTICS (USA)
ADVANCED HOLOGRAPHIC LABS (USA)
AG ELECfRO-OPTICS LTD. (United Kingdom)
AMERICAN HOLOGRAPHIC (USA)
AMITY PHOTONICS CO. (USA.)
ASAHI GLASS CO. (Japan)
ApA OPTICS, INC. (USA.)
BRITISH AEROSPACE (United Kingdom)
CENTRAL GLASS CO., LTD. (Japan)
CHROMAGEM INC. (USA)
CISE SPA TECHNOLOGIE INNOVATIVE. (Italy)
CONTINENTAL OPTICAL (USA)
CONTROL OPTICS (USA)
CORION CORP. (USA)
COULTER OPTICAL COMPANY (USA.)
CSI. (United Kingdom.)
CZECHOSLOVAK ACADEMY OF SCIENCE (Czechoslovakia)

DAIMLER BENZ AG. (Germany)
DATASIGHTS LTD. (United Kingdom.)
DAVIN OPTICAL LTD. (United Kingdom)
DELL OPTICS COMPANY, INC (USA.)
E.I. DUPONT DE NEMOURS & CO., IN (USA.)
EALING ELECTRO-OPTICS (UK) (United Kingdom)
EALING ELECTRO-OPTICS INC. (USA)
ELECTRO OPTICAL INDUSTRIES, INC. (USA)
ELECTRO OPTICS DEVELOPMENTS LTD. (United Kingdom.)
EXPANDED OPTICS LIMITED (United Kingdom)
FRESNEL TECHNOLOGIES INC (USA.)
FUn PHOTO OPTICAL CO. (Japan.)
G.M. VACUUM COATING LAB (USA)
GALVOPTICS LTD. (United Kingdom.)
HOLO 3 (France)
HOLO-OR LTD (Israel)
HOLOGRAPHIC OPTICS INC (USA)
HOLOGRAPHIC STUDIOS (USA)
HOLOGRAPHY INSTITUTE (USA)
HOLOGRAPHY ISRAEL (Israel)
HOLOGRAPHY NEWS (United Kingdom)
HOLOGRAPHY WORLD CENTER (USA)
HOLOMEDIA AB/HOLOGRAM MUSEUM. (Sweden)
HOLOPRINT ROSOWSKI (Germany)
HOLOPUBUC UNBEHAUN (Germany)
HOWARD SMITH PRECISION OPTICS (Unted Kingdom)
HUGHES AIRCRAFT CO.- (USA)
HYOGO PREFECTUAL MUSEUM Japan
IBM ALMADEN RESEARCH CENTER (USA)
ICI AMERICAS (USA)
ILLINOIS INST OF TECHNOLOGY USA
IMEOOE TECHNOLOGY (USA)
IMPERIAL COLLEGE OF SCIENCE (United Kingdom)
INFOTECH INTERNATIONAL (USA)
INFRARED OPTICAL PRODUCTS, INC. (USA)
INRAD, INC (USA)
INSTITUTE OF NUCLEAR PHYSICS (Russia)
INSTITUTE OF OPTICAL SCIENCE (Taiwan)
INSTITUTE OF PHYSICS (Russia)
INSTITUTE OF PLASMA PHYSICS (Poland)
INTERNATIONAL HOLOGRAM (United Kingdom)
ISAST/LEONAROO (USA)
JOURNAL OF LASER APPUCATIONS (USA)
KAISER OPTICAL SYSTEMS (USA)
KAROLINSKA INSTITUTET (Sweden)
KEIO UNIVERSITY (Japan)
KENDALL HYDE LTD. (United Kingdom)
KREISCHER OPTICS, LTD. (USA)

L.I.R. E.R.A. (France)
LAMBDAffEN OPTICS (USA)
LASER OPTICS, INC. (USA)
MMG MINNAHUETTE MASCHINELLE GLAS (Germany)
MODERN OPTICS (V-Tech) (USA)
NIPPONDENSO CO., LTD. (Japan)
ODHNER HOLOGRAPHICS (USA)
OPTICAL IMAGES (USA.)
OPTICAL SURFACES LTD. (United Kingdom)
OPTICAL WORKS LTD. (United Kingdom
OPTICS PLUS INC (USA)
PENNSYLVANIA PULP & PAPER (USA)
PHYSICAL OPTICS CORP (USA)
RALCON (USA)
RICHTER ENTERPRISES (USA)
ROCHESTER PHOTONICS CORPORATION.
(USA)
SPECTROGON (USA) (USA)
SPECTROGON AB (Sweden)
UK OPTICAL SUPPLIES -United Kingdom
UNIVERSITH ERLANGEN - NURNBERG (Germany)
VINTEN ELECTRO OPTICS LTD (United Kingdom)
Businesses Involved In NonDestructive Testing:
ABBOTT LABORATORIES (USA)
ADVANCED ENVIRONMENTAL RESEARCH (USA)
AEROSPATIALE (France)
AEROSPATIALE. (France)
ALABAMA A&M UNIVERSITY (USA)
BEITING NORMAL UNIVERSITY China
BIAS (Germany)
BRmSH AEROSPACE PLC. (United Kingdom)
BROOKHAVEN NATIONAL LAB (USA)
CHERNOVTSY STATE UNIVERSITY (Russia)
CISE SPA TECHNO LOGIE (Italy)
CITRoEN INDUSTRIE. (France)
CRANFIELD INSTITUTE OF TECHNOLOGY
(United Kingdom.)
CZECHOSLOVAK ACADEMY OF SCIENCE (Czechoslovakia)
DAIMLER BENZAG. (Germany)
DUSTON HOLOGRAPHIC SERVICES (Canada)
ELECTRO OPTIC CONSULTING SERVICE (USA)
FORD RESEARCH STAFF (USA.)
HOLO 3 (France)
HOLOFLEX COMPANY (USA)
HUGHES AIRCRAFT CO. (USA)
ING.-AGENTUR FUR NEUE (Germany)
LAB CENTRAL DES PONTS ET CHAUSSE (France.)
LABOR DR. STEINBICHLER (Germany)
LABORATORY SOETE (Belgium)
LAMBDA ANALYTICAL LABS (USA.)

LASER APPLICATIONS, INC. (USA)
LASER TECHNOLOGY, INC. (USA)
LASERMET LIMITED (United Kingdom)
LAZAP INC (USA)
LITTON SYSTEMS CANADA (Canada)
LULEA UNIVERSITY OF TECHNOLOGY D (Sweden)
MAZDA MOTOR CORP. (Japan)
METROLASER. (USA)
MICRAUDEL. (France)
MITSUBISHI HEAVY INDUSTRIES LTD. (Japan)
McMAHAN ELECfRO-OPTIC (USA)
NATIONAL PHYSICAL LABORATORY (United Kingdom.)
NEW YORK INST OF TECH. (USA.)
NORTH DANCER HOLOGRAPHIC (USA
NORTHWESTERN UNIVERSITY (USA.)
NOV ATOR RESEARCH CENTER (Russia)
PHOTONICS SYSTEMS LABORATORY (France)
ROTIENKOLBER HOLO-SYSTEM GmbH (Germany)
ROYAL INSTITUTE OF TECHNOLOGY (Sweden)
RUTHERFORD & APPLETON LABS. (United Kingdom)
SAAB-SCANIA (Sweden)
SANDIA NATIONAL LABORATORIES (USA)
SPECfRON DEVELOPMENT LAB(USA)
STOLTZ AG (Switzerland)
SWISS FEDERAL INSTITUTE OF TECHN (Switzerland)
ULTRAFINE. (United Kingdom)
UNIVERSIDADE DO PORTO (Portugal)
UNIVERSITE DE NEUCHATEL (Switzerland)
UNIVERSITY OF ABERDEEN (United Kingdom)
UNIVERSITY OF ALABAMA (USA)
UNIVERSITY OF ARIZONA (USA)
UNIVERSITY OF MICHIGAN (USA)
UNIVERSITY OF ROCHESTER (USA)
UNIVERSITY OF SOUTHERN CALIF. (USA)
UNIVERSITY OF STRATHCLYDE (United Kingdom)
UNIVERSITY OF STUTIGART (Germany)
UNIVERSITY OF TOKYO (Japan)
UNIVERSITY OF WISCONSIN (USA)
UNIVERSITY OF ZAGREB (Yugoslavia)
UNIVERSITH ERLANGEN - NURNBERG (Germany)
VOLKSWAGEN AG (Germany)
VOLVO-FLYGMOTOR (Sweden)
WORCESTER POLYTECH. INST (USA)
WYKO Corp. (USA)

Places To See Holograms:

AMAZING IMAGES (USA)
AMAZING WORLD OF HOLOGRAMS. (United Kingdom)
ART INSTITUTE OF CHICAGO (USA)
ART, SCIENCE & TECHNOLOGY (USA.)
ARTBRIDGE MANAGEMENT (Germany)
ARTIGLIOGRAPHY CO. (USA)
ATLANTA GALLERY OF HOLO. (USA)
CENTRE D'ART (Canada)
DREAM IMAGES. (Germany)
ELUSIVE IMAGE. (USA)
FOUNDATION IDEECENTRUM. (Netherlands)
GALLERIE ILLUSORIA (Switzerland.)
HELLENIC INSTITUTE OF HOLOGRAPHY (Greece)
HOLOGRAMM WERKSTATI & GALERIE (Switzerland)
HOLOGRAPHIC INDUSTRIES, INC -USA
HOLOGRAPHIC STUDIOS (USA)
HOLOGRAPHY CENTER OF AUSTRIA .. (Austria.)
HOLOGRAPHY ISRAEL (Israel)
HOLOGRAPHY WORLD CENTER (USA)
HOLOLASER GALLERY (U.A.E.)
HOLOMEDIA AB/HOLOGRAM MUSEUM. (Sweden)
HOLOS ART GALERIE (Switzerland)
HYOGO PREFECfUAL MUSEUM Japan
INTERFERENS HOLOGRAFI D.A. (Norway)
KARAS STUDIOS S.L. (Spain)
LABOR FuR HOLOGRAFIE (Germany)
LASER AFFILIATES (USA)
LASER INTERNATIONAL (United Kingdom)
LIGHT WAVE GALLERY (USA.)
LONE STAR ILLUSIONS (USA)
MAGIC LASER (France) (France)
MAGIC LASER (SPAIN) (Spain)
MAGIC LIGHT HOLOGRAFIE GALLERIE (Germany)
MUNDAY SPATIAL IMAGING. (United Kingdom)
MUSEUM FuR HOLOGRAPHIE Germany
MUSEUM OF HOLOGRAPHY/CHICAGO. (USA)
MUSeE DE L'HOLOGRAPHIE (France)
NEW YORK HALL OF SCIENCE (USA.)
ONTARIO SCIENCE CENTRE (Canada.)
SCHAFFNER GALLERY (USA)
SCIENTIFIC COUNCIL ON EXHIBmON (Russia)
SEMICON AUSTRIA (Austria)
SHANDONG ACADEMY OF SCIENCES (China)
STARLIGHT HOLOGRAPHIC (Canada)
THREE-D LIGHT GALLERY (USA)
TREND (Yugoslavia)
WHITE LIGHT WORKS, INC. (USA)
WHOLE PICfURE (USA)

Holography Artists:

3D VISION (Germany)
3D1 (United Kingdom)
ACME HOLOGRAPHY (USA)
ADVANCED IMAGING TECHNOLOGIES (United Kingdom)
AKS HOLOGRAPHIE-GALERIE GmbH. (Germany)
AP HOLOGRAFIKA STUDIO (Hungary)
ARBEITSKREIS HOLOGRAFIE BV (Germany)
ARCHEOZOIC INCORPORATED. (USA)
ARCHITECfURAL GLASS & HOLOGRAPH (United Kingdom)
ARMIN KLIX HOLOGRAPHIE Germany
ART & DESIGN GRAPHICS (Russia)
ART FREUND HOLOGRAPHY (USA)
ART INSTITUTE OF CHICAGO (USA)
ARTBRIDGE MANAGEMENT (Germany)
ARTIGLIOGRAPHY CO. (USA)
ARTKITEK(USA)
ASCOT LASER PICfURE STUDIO (United Kingdom)
ATELIER HOLOGRAPHIQUE DE PARIS (France)
ATLANTA GALLERY OF HOLOGRAPHY (USA)
AUSTRALIAN HOLOGRAPHICS PTY LTD. (Australia)
BARR & STROUD (United Kingdom)
BENYON, MARGARET United Kingdom
BERKHOUT, RUDIE (USA)
BOB MADER PHOTOGRAPHY. (USA)
BOOTH,ROBERTA(USA)
BOYD, PATRICK (United Kingdom)
BURNS HOLOGRAPHICS LTD (USA)
BYRNE,KENNETH(USA)
CAMBRIDGE STEREOGRAPHICS GROUP. (USA)
CASDIN-SILVER HOLOGRAPHY. (USA)
CENTRE D'ART (Canada)
CHERRY OPTICAL HOLOGRAPHY. (USA.)
COSSETIE, MARIE ANDREE (Canada)
CROSS, LLOYD (USA)
DAVID SCHMIDT HOLOGRAPHY USA.
DEEM,REBECCA(USA)
DEEP SPACE HOLOGRAPHICS (Canada)
DEUTSCHE GESELSHAFT (Germany)
DIALECTIC A AB (Sweden.)
DIAURES S.A. HOLOGRAPHIC DIVISIO (Italy)
DREAM IMAGES. (Germany)
DUSTON HOLOGRAPHIC SVS.(Canada)
EIF PRODUCTIONS (France.)
ED WESLY HOLOGRAPHY. (USA)
ELECfRO OPTIC CONSULTING (USA)
FLATIRON STUDIO (USA)
FORNARI, DAVID (USA)
FRINGE RESEARCH HOLO. (Canada.)
FTI JOFFE (Russia)

GENERALFEROE(USA)
GERALD MARKS STUDIO. (USA.)
GRAY SCALE STUDIOS LTD. (USA)
HARRIS, NICK (USA)
HELLENIC INST. OF HOLOGRAPHY (Greece)
IDGH TECH NETWORK (Sweden.)
HM-Holographie (Germany)
HOECHST CELANESE CORP (USA)
HOLAGE (USA.)
HOLICON CORPORATION. (USA)
HOLO GMBH HOLOGRAFIELABOR (Germany)
HOLO-DIMENSIONS INC (Canada)
HOLO-IMAGES, INC. (USA)
HOLO-LASER. (France)
HOLO-SERVICE. (Switzerland)
HOLO-SERVICE.FRIES (Switzerland)
HOLO-SOURCE CORPORATION (USA)
HOLO-SPECTRA (USA)
HOLOCOM HOLOGRAPHIE (France.)
HOLOCRAFT INTERNATIONAL (USA)
HOLOFAR LAB (SRL) (Italy)
HOLOGRAMM WERKSTATT & GALERIE (Switzerland)
HOLOGRAMS 3D (United Kingdom)
HOLOGRAMS FANTASTIC & ILLUSIONS (Australia)
HOLOGRAPHC MARKEITNG (USA)
HOLOGRAPIDC CONCEPTS (USA)
HOLOGRAPIDC DIMENSIONS, (USA)
HOLOGRAPHIC IMAGES INC. (USA)
HOLOGRAPIDC STUDIO (Canada)
HOLOGRAPIDC STUDIOS (USA)
HOLOGRAPIDCS (UK) LTD. (United Kingdom)
HOLOGRAPIDCS NORTH INC. (USA)
HOLOGRAPHIE LABOR (Germany)
HOLOGRAPHY CENTER of AUSTRIA .. (Austria.)
HOLOGRAPHY INSTITUTE (USA)
HOLOGRAPHY ISRAEL (Israel)
HOLOMAT (USA)
HOLOMEDIA AB/HOLOGRAM MUSEUM. (Sweden)
HOLOMORPH VISUALS, INC. (Canada)
HOLOPRESS KG (Germany)
HOLOPRODUCTION. (France)
HOLOTEC (Germany)
HOLOTEC BIRENHEIDE (Germany)

HOLOVISION AB. (Sweden)
IMEDGE TECHNOLOGY (USA)
INTERFERENS HOLOGRAFI (Norway)
ISHII, MS. SETSUKO (Japan)
JAYCO HOLO. (United Kingdom)
K.C. BROWN HOLO. (United Kingdom)
KAC, EDUARDO (USA)
KAUFFMAN, JOHN (USA)
KYOTO TECHNICAL UNIV. (Japan)
L.A.S.E.R. CO. (USA)
LABOR FUR HOLOGRAFIE (Germany)
LAMINEX/HIGH TECH UK (United Kingdom)
LASART LTD. (USA)
LASER LIGHT DESIGNS (USA)
LASER LIGHT IMAGE. (United Kingdom)
LASER LIGHT LTD. (USA)
LASERFILM ECKHARD KNUTH (Germany)
LAS ERION HANDELS GMBH (Germany)
LASERMEDIA. (USA.)
LASERSMITH, INC. (USA.)
LASERWORKS (USA)
LASIRIS INC. (Canada)
LAZA HOLOGRAMS (United Kingdom)
LAZART HOLOGRAPHICS. (Australia.)
LES PRODUCTIONS HOLOLAB! Canada
LEVINE, CHRIS (United Kingdom)
LIBERATO, PAUL (USA)
LIGHT ENGINEERING (United Kingdom)
LINDA LAW HOLOGRAPIDCS (USA)
MACSHANE HOLOGRAPHY (USA.)
MAN ENVIRONMENT, INC. (USA.)
MARTINS SON ELEKTRONIK . (Sweden)
MC CORMACK, SHARON (USA)
MGD PRODUCTIONS (Canada)
MGM CONVERTERS INC (USA)
MILLER,PETER(USA)
MOELLER WEDEL OPTISCHE WERK (Germany)
MULTI-PURPOSE HOLOGRAMS(France)
MUNDAY SPATIAL IMAGING. (United Kingdom)
MUSEUM OF HOLOGRAPHYjCHICAGO(USA)
MU's LASER WORKS (Canada)
NAIMARK, MICHAEL (USA)
NATIONAL HOLO. STUDIOS (USA)
NEW DIMENSION HOLOGRAPHICS. (Australia)
NEW YORK HALL OF SCIENCE (USA.)

NEW YORK HOLOGRAPIDC LABS (USA)
NORTH AMERICAN HOLOGRAPIDCS (USA.)
NORTH DANCER HOLOGRAPIDC -USA
NORTHERN LIGHTWORKS (USA)
OIE RESEARCH (USA)
OPTICAL LABORATORY (France)
ORIEL CORPORATION (USA)
ORIEL SCIENTIFIC (United Kingdom)
PARKER, JULIE WALKER (USA)
PHOENIX HOLOGRAMS (Germany)
PINK, PATTY (USA)
POINT OF VIEW DIMENSIONS (USA)
POINT SOURCE PRODUCTIONS (USA)
POLYMER HOLOGRAPIDCS INC. (USA)
RANDAZZO,DEAN(USA)
RANDY JAMES HOLOGRAPHY (USA)
RAVEN HOLO. (United Kingdom)
RED BEAM, INC. (USA)
REEL IMAGE (USA)
REGAL PRESS INC (USA)
RICHARD BRUCK (USA.)
ROBERT SHERWOOD HOLO DESIGN (USA)
ROY, JEAN (Canada)
ROYAL PHOTOGRAPIDC SOCIETY (United Kingdom)
RUEY-TUNG, MISS. HUNG (Japan)
SOCIETA OLOGRAFICA IT ALIA (Italy)
SOPIDA UNIVERSITV (Japan)
SPECAC LTD (United Kingdom)
STALLARD,PENN(USA)
STEPHENS, ANAIT (USA)
STUDIO CREAIUS (USA)
SUPERIOR TECHNOLOGY IMPLEMEN-TATI (Denmark)
SWEDE HOLOPRINT (Sweden)
THREE-D LIGHT GALLERY (USA)
TJING LING INDUSTRIAL RESEARCH. (Taiwan)
TRIDIMENSIONALE HOLOGRAMAS (Spain)
ULTIMATE IMAGE (Australia)
UNTERSEHER & ASSOCIATES (USA)
WAVE MECHANICS (USA)
WAVEFRONT TECHNOLOGIES (USA)
WHITE LIGHT WORKS, INC. (USA)
WHOLE HOGRAPHY (USA)
WORLD ART PROJECT (USA)

SECTION 6

Bibliography
Glossary
Index

15

Bibliography

For further reading on holography and its many applications, we provide this bibliography. It has been divided into ten categories: artistic holography; general holographic applications; exhibition catalogues; general dissertations ; industrial holography; industrial dissertations; holography marketing; periodicals,; non-English holography material, and laser-related titles.

If you are aware of any other sources which should be included, please forward all information to Holography Market Place.

Artistic Holography

Allen, Judy. Lasers and Holograms. England: Puffin Books, UK, 1985.

Anderson, John. Holography. Tempe: Art Dept., Arizona State University, 1979. Series title: Northlight; no.11.

Barrett, N. S. Lasers & Holograms. Boston: Watta, 1985. Series Title: Picture Library.

Berley, Lawrence F. Holographic mind, holographic vision: a new theory of vision in art and physics. 1st ed. Bensalem, PA: Lakstun Press, 1980.

Burkig, Valerie. Photonics: The New Science of Light. [n.I.]: Enslow Publishers, 1986.

Caulfield, H. John; (et al). Holography Works. New York: Museum of Holography (NYC), 1984.

Centerbeam. Otto Piene and Elizabeth Goldring, eds. Introduction by Lawrence Alloway. Cambridge, MA: Center for Advanced Visual Studies, Massachusetts Institute of Technology, 1980.

Dowbenko, George. Homegrown holography. Garden City, N.Y.: Amphoto, 1978.

Easy Way to Make Reflection Holograms (no author). Embee Press, 1986.

Falk, David R. (et alia). Seeing the Light: Optics in Nature, Photography Color, Vision, & Holography. New York: Wiley, 1985.

Finch College, New York. Museum of Art. Contemporary Study Wing. "N dimensional space". Prepared by Ted McBurnett. Introd. by Elayne H. Varian. New York: 1970.

Furst, Anton. et alia. Light fantastic. 2d ed. London: Bergstrom & Boyle Books, 1977. Graphics in motion: from the special effects film to holographics. John Halas, ed. New York: Van Nostrand Reinhold, 1984.

Griffiths, John. Lasers & Holograms. New York: Silver Publishers, 1983. Series Title: Exploration & Discovery Series.

Heckman, Philip. The Magic of Holography. New York, Macmillan & Co., 1986.

The Holographic paradigm and other paradoxes : exploring the leading edge of science. Ken Wilber, ed. 1st ed. Boulder: Shambhala, 1982.

Holographic recording materials. H. M. Smith, ed. Contributions by R. A. Bartolini... [et al.]. Berlin: Springer-Verlag, 1977. Series

title: Topics in applied physics; v. 20.

Holography Redefined: Thresholds. Barilleaux, Rene P., Editor. New York: Museum of Holography, 1984.

International Exhibition of Holography. Jeong, Tung and Michael Croydon, Editors. Lake Forest, IL: Lake Forest College Press, 1982.

(Second) International Exhibition of Holography. Jeong, Tung, Editor.Lake Forest, IL: Lake Forest College Press, 1985.

Jeong, Tung H. Display Holography: Proceedings of the International Symposium, 1982;

Vol. 1. Lake Forest, IL: Lake Forest College Press, 1983.

_ _ Display Holography: Proceedings of the International Symposium, 1982; Vol. II. Lake Forest, IL: Lake Forest College Press, 1986.

Kallard, Thomas. Laser Art & Optical Transforms. New York: Optosonic Press, 1979.

Kasper, Joseph Emil and Steven Feller. The complete book of holograms: how they work and how to make them. New York: Wiley, 1987. Series title: The Wiley science editions.

_ _ _ . The hologram book. Englewood Cliffs, N.J.: Prentice-Hall, 1985.

Lancaster, Ian M. (et alia). The Holographic Instant: Pulsed Laser Holography. New York: Museum of Holography, 1987.

Light vistas light visions. Sponsored by the

174 Holography Market Place -Fourth Edition Department of Art, St. Mary's College. Notre Dame, IN: St. Mary's College, 1983.

Lucie-Smith, Edward. Art In The Seventies.Ith- aca, NY: Cornell University Press, 1980. Contrib: R. Berkhout, H. Casdin-Silver, P.Claudius.

Neumann, Don Barker. "The effect of scene motion on holography". Columbus, OH: dissertation, 1967.

Palin, Michael. The Mirrorstone: A Ghost Story with Holograms. New York: Knopf,1986.
Walton, Paul. Space-light: a holography and laser spectacular. London: Routledge & Kegsn Paul,1982.

Artistic Holography: Exhibit Catalogues And Miscellany
Bienal International de sao Paulo, Catalogo Geral. sao Paulo, Brasil: Fundacao Bienal de Sao Paulo, 1985. See ·Entre a Ciencia EA Ficcao", pp167-197, re: holographers M. Baumstein, H. Casdin-Silver, J.W.Garcla.

Alice in the Light World. 'Ibkyo, Japan: The Ashai Shimbun, 1978.

Art Trsnsition:. Cambridge, MA: MIT Center for Advanced Visual StudieslUniversity Film Study Center, 1975. See H. Casdin-Silver ·Holography...• pp30-32.

Critic's Choice: The Craft of Art; Peter Moore's Liverpool Project 5. Liverpool, England: November 3, 1979. See E. Lucie-Smith, "New Attitudes, New Materials, New Techniques".

_ _ _ _ ,. Liverpool, England:Walker Art Gallery, 1979. No. 84-110. See W.D.L. Scobie, ·Arts Review".

Electra 83. Paris: Les Amis du Musee d'Art de la Ville de Paris, 1983. See F.Popper, ·Electricity and Electronics in the Art of the 20th century", pp 46-50.

Expansion. Internationale Biennale fur Graphik und Visuelle Kunst. Horst Gerhard Haberl, Generalsekretar. Wien, Austria: Inter- nationale Biennale fLir Graphik und Visuelle Kunst, 1979. See O.Peine "MIT-Center for Advanced Visual Studies', ·Sky Events" pp 232- 239; E_ Goldring ·Documentation room", ·Cen- terbeam" p_ 232; H_Casdin-Silver ·Holography, a holographic environment" p_ 234; G. Kepes ·Art of the Environment' p. 99.

Fantasy of Holography. Tokyo, Japan: Seibu Museum of Art, 1976. Itsuo Sakane, ed. Contr: Shuntoro Tankawa, Junpei Tsujiuchi.

Harriet Casdin-Silver Holography. New York: Museum of Holography, 1977. First one person exhibition at the Museum

High Technology and Art 1986. 'Ibkyo, Japan: 'Ibkyo Slrimbun & Nagoya Shimbun, with Asso- ciates of Art and Technology, Japan ,

'Holography redefined', Thresholds'. Harriet Casdin-Silver with Dov. Eylath. New York: Museum of Holography, 1984. Group exhibition.

Inter. Quebec, Canada: Les Editons Intervention, Printemps 86, no,31, 1986. See E. Shapiro,

·Art, Perception et Holographie" p. 32; L.Heaton, 'Tire D'une Entrevue avec Harriet Casdin-Silver" pp32-3.

Images in Time and Space. Ottawa, Ontario, Canada: Association of Science and Technology Inc, 1987. Trsvelling exhibit.

International Holography. London, England: The Photographers' Gallery, 1980.

Light and Substance. New Mexico: University of New Mexico Art Museum, 1973-75. History of photography, holography by S. Benton, H.Cas- din-Silver. Van deren Coke, org.

MultiMedia Exhibition. Kansas City, MO: Nelson Gallery at Atkins Museum of Fine Arts, 1970. See ·Holography by H,Casdin-Silver".

Otto Peine und CAVS: 20th Annivell!ary CAVS. Karlsruhe, West Germany: Badischer Kunstv- erein, 1988. See H. Casdin-Silver, A. Cheji, D. Jung, J.Powell.

Sky art conference-'81. Cambridge, MA: MIT Center for Advanced Visual Studies, 1981. L. Burgess, E. Goldring, B. Kracke, O. Piene, eds. See H.Casdin-Silver, ·Sky work: solar-tracked hologram series" p.49,

Sky art conference '83. Cambridge, MA: MIT Center for Advanced Visual Studies with der Landeshaupstadt Munchen der BMW AG und der Digital Equipment GmbH, 1983.

General Holography Information
Abramson, Nils H. The making and evaluation of holograms. London: Academic Press, 1981.

Advances in Holography. Farhat, Nabil H.-Editor. New York: Marcel Dekker,1975.Vol. 1.

Advances in Holography. Farhat, Nabil H.-Editor. New York: Marcel Dekker,1976,Vol. 2.

Advances in Holography. Farhat, Nabil H.-Editor. New York: Marcel Dekker,1976.Vol. 3.

Applications of Holography: January 21-23, 1985, Los Angeles, California. Lloyd Huff, chair, ed. Bellingham, WA: SPIE--the International Society for Optical Engineering, 1985, Series title: Proceedings of SPIE--the International Society for Optical Engineering; v. 523.
Berner, Jeff. The holography book. New York: Avon Books,1980.

Berry, Michael V, Diffraction of Light by Ultra- sound. New York: Academic Press,1967.

Cadig Liaison Centre_ Reference Library. A compendium of Cadig bibliographies: Metrication, Fluidics, Explosive techniques in engineering, Holography, Carbon fibres. Coventry (Warwickshire): Cadig Liaison Centre, 1970.

Cathey, W. Thomas. Optical information processing and holography. New York:Wiley,1974. Series title: Wiley series in pure and applied optics.

Caulfield, H. J. and Sun Lu. The applications of holography. New York, Wiley-Interscience, 1970. Series title: Wiley series in pure and applied optics,

Chambers, R.P. and J.S. Courtney-Pratt. Bibli- ography on Holograms. New York: Bell Tele- phone Laboratories,1966.

Chuguy, Y a V. and N. T. Kolesova_ Bibliography on holography, 1971-1972. Trsnslated from Russian. London: Scientific Information Con- sultants Ltd, 1976.

Coherent optical processing: seminar, August 21-22, 1974, San Diego, CA_ Palos Verdes Estates, CA: Society of Photo-optical Instrumentation Engineers, 1975. Series title: Society of Photo-optical Instrumentation Engineers Seminar proceedings; v. 52.

Coherent optics in mapping: tutorial seminar and technology utilization program, March 27-29, 1974, Rochester, N.Y. N. Balasubramanian, Robert D. Leighty, eds. Jointly sponsored by American Society of...Palos Verdes Estates, CA: SPIE,1974. Series title: Society of Photo-optical Instrumentation Engineers Proceedings; v. 45.

Collier, Robert Jacob, C. B. Burckhardt and L,H. Lin, Optical holography. New York: Academic Press, 1971.

Collings, Neil. Optical pattern recognition using holographic techniques. Wokingham, England; Reading, MA. : Addison-Wesley Pub. Co., 1988. Series title: Electronic systems engi- neering series.

Collings, Neil. Optical Pattern Proceedings. Symposium on High-Power Lasers and Optical Computing, OEILASE '90, 14-19 January 1990, Los Angeles, California. SPIE--the Interna- tional Society of Engineering; v. 1211.

Computer-generated holography II: 11-12 Jan- uary 1988, Los Angeles, CA. Sing H. Lee, chair! editor; the International Society for Optical Engineering; cooperating... Bellingham, WA., USA: SPIE--the International Society for Opti- cal Engi neeri ng, 1988.Series title: Proceedings of SPIE--the International Society for Optical Engineering; v. 884.

Conference on Fourier Optics, Lasers and Holography, Mysore, 1971. Proceedings of the Conference on Fourier Optics, Lasers and Holography, Mysore, November 11-15, 1971. Madras, India: Institute of Mathematical Sciences, 1971. Series title: Matscience report; 77,

Conference on Holography and Optical Filter- ing, Marshall Space Flight Center, 1971: Holography and optical filtering; proceedings, Washington, Scientific and Technical Information Office, National Aeronautics and Space Administration.

Conference on Holography and Optical Filter- ing,1972: Marshall Space Flight Center. Holog- raphy and optical filtering; proceedings. Washington, Scientific and Technical Informa- tion Office, National Aeronautics and Space Administration,1973_ Series title: United States National Aeronautics and Space Administration NASA SP -299.

Defense Documentation Center (U.S_) Holography: a DOC bibliography, January 1970-Sep- tember 1972. Alexandria, VA: Defense Documentation Center, Defense Supply Agen- cy, 1973.

Denisiuk, IU. N. Fundamentals of holography. Translated from the Russian by Alexander Chu- barov. Rev. from the 1978 Russian ed. Moscow: Mir,1984.

DeVelis, John B. and George O. Reynolds. Theory and applications of holography. Reading, MA: Addison-Wesley Pub. Co., 1967.

Developments in holography: seminar-indepth; proceedings. Brian J. Thompson and John B. DeVelis,eds. Redondo Beach, CA: Society of Photo-optical Instrumentation Engineers, 1971. Series title: Society of Photo-optical Instrumentation Engineers S.P .I.E.seminar proceedings, v. 25.

Dudley, David D. Holography; a survey. Washington, DC: Technology Utilization Office, National Aeronautics and Space Administration,1973. Series title: NASA SP-5118.

Eichert, Edwin S. and Alan H. Frey; Randomline, Inc. Holography in driver education, training, testing, and research. Washington, DC : National Highway Traffic Safety Administration,1978. Series title: United States. National Highway Traffic Safety Administration Report; no. DOT HS-803 035.

Engineering Applications of Holography Sympo- sium,1972: Los Angeles: Proceedings. Redondo Beach, CA: Society ofPhoto-optical Instrumentation Engineers, 1973.

The Engineering uses of coherent optics: proceed- ings and edited discussion ofa conference held at the University of Strathclyde, Glasgow, 8-11 April,1975. Organised by the University, in asso- ciation with the... Cambridge, Eng.: Cambridge University Press, 1976.

The Engineering uses of holography. Robertson, Elliot R. and James M. Harvey, eds. Cambridge, England: Cambridge University Press, 1970.

European Hybrid Spectrometer Workshop on Holography and High-Resolution Techniques,1981: Strasbourg, France. Photonics applied to nuclear physics, 1. European Hybrid Spectrometer Workshop on Holography and High-Resolution Techniques, Strasbourg, Council of Europe, 9-12 November 1981. Geneva: Euro- pean Organization for Nuclear Research, 1982. Series title: CERN (Series) ; 82-01.

An External Interface for Processing 3-D Holo- graphic and X-ray Images: Werner Juptner, Tho- mas Kreis (eds.). Berlin: Springer-Verlag,1989.

Fiber Diffraction Methods. French, Alfred D. and Kenn Gardner, Editors. Series Title: ACS Sympo- sium Ser.,; No. 141. New York: American Chemical Society,1980.

Firth, Ian Mason. Holography and computer gen- erated holograms. London, Mills and Boon, 1972. Series title: M & B monograph EElll.

Francon, M. Holography. Expanded and revised from the French edition. Translated by Grace Marmor Spruch. New York: Academic Press, 1974.

Handbook of Optical holography. H. J. Caulfield,ed. Contributors, Gilbert April... let aLl New York : Academic Press, 1979.

Hariharan, P. Optical holography : principles, techniques, and applications. Cambridge: Cambridge University Press,l983. Series title: Cambridge monographs on physics.

Hildebrand, B. P. and B. B. Brenden. An introduction to acoustical holography. New York: Plenum Press, 1972.

"Holographic detection of intraocular pathology in the presence of cataracts: final report". By George O. Reynolds... let al.). Burlington, MA: Technical Operations, 1974.

Holographic optics: design and applications: 13-14 January 1988, Los Angeles, CA. Ivan Cindrich, ChairlEditor ; SPIE--The International Society for Optical Engineering; cooperating... Bellingham, WA., USA: SPIE--The International Society for Optical Engineering, 1988.Series title: Proceedings of SPIE--The International Society for Optical Engineering; v. 883.

Holographic optics: optically and computer gen- erated: 19-20 January 1989, Los Angeles, CA . Ivan N. Cindrich, Sing H. Lee chairs/editors; sponsored by SPIE--The International Society for Optical... Bellingham, WA., USA : SPIE- The International Society for Optical Engineering, c1989. Series title: Proceedings of SPIE--the International Society for Optical Engineering; v. 1052.

Holography. Redondo Beach, CA: Society of Photo-optical Instrumentation Engineers, 1968. Series title: Society ofPhoto-opticalln- strumenta- tion Engineers S.P.I.E. seminar proceedings, v. 15.

Holography Applications. Wang. Editor. Bell- ing- ham, WA: SPIE,1986.

Holography: Critical Reviews. Huff, L. , Editor. Bellingham, WA: SPIE, 1985.

Holography: January 24-25, 1985, Los Angeles, CA. Lloyd Huff, chair,ed. Bellingham, WA: SPIE-the International Society for Optical Engineering,1985. Series title: Proceedings of SPIE--the International Society for Optical Engineering; v. 532. Series title: SPIE critical reviews oftechnol- ogy series; 12th.

Holography; seminar-in-depth, May, 1968, San Francisco CA. B.G.Ponseggi and Brian J. Thomp- son,eds. Redondo Beach, CA: Society of Photo- optical Instrumentation Engineers, 1972. Series title: Society of Photo-optical Instrumentation Engineers Proceedings, v.15.

Holography techniques and applications: EC01, 19-21 September 1988, Hamburg, Federal Republic of Germany / Werner P.O. Juptner, chair/editor; EPS--European Physical Society, Europtica--the... Belling- ham, WA. : SPIE--the International Society for Optical Engineering, 1989. Series title: Proceedings ofSPIE--the Inter- national Society for Optical Engineering ; v. 1026.

I. Aroslavskii, L. P. and N. S. Merzlyakov. Meth- ods of digital holography. Translated from Rus- sian by Dave Parsons. New York: Consultants Bureau, 1980.

International Commission for Optics. Congress,lOth : 1975 : Prague, Czechoslovakia. Recent advances in optical physicsz: proceedings of the Tenth Congress of the International Commission for Optics, August 25-29, 1975, Prague, Czechoslovakia. Bedrich Havelks and Jan Bla- bla,eds. Olomouc: Palacky University; Prague: Society of Czechoslovak Mathematicians and Physicists, 1976.

International Conference on Applications of Holography and Optical Data Processing, 1976: Jerusalem, Israel. Applications of holography and optical data processing: proceedings of the international conference, Jerusalem, August 23- 26, 1976. E. Marom, A. A. Friesem, and E. Wiener-Avnear, eds. 1st ed. Oxford: Pergamon Press, 1977.

International Conference on Computer-generated Holography, 1983: San Diego, CA. International Conference on Computer-generated Holography, August 25-26,1983, San Diego, CA: proceedings. Sing H. Lee, chair, ed. Bellingham, WA: SPIE--The International Society for Optical Engineering,1983. Series title: Proceedings of SPIE--the International Society for Optical Engi- neering; v. 437.

International Conference on Computer-generated Holography: 2nd: 1988: Los Angeles, CA. Computer-generated Holography II: 11-12 January 1988, Los Angeles,California, [proceedings) . Sing H. Lee, chair/ed. Sponsored by SPIE--The International Society for Optical Engineering ;... Bellingham, WA: SPIE--The International Society for Optical Engineering, 1988. Series title: Proceedings of SPIE--the International Society for Optical Engineering; v. 884.

International Conference on Holographic Systems, Components and Applications,1987: Churchill College. Holographic systems, compo- nents and applications: Churchill College, Cam- bridge, 10th-12th September, 1987. London: Institution of Electronic and Radio Engineers, 1987. Series title: Publication / Institution of Electronic and Radio Engineers; no. 76. International Conference on Holographic Sys- tems, Components, and Applications. 2nd:1989: University of Bath.

International Conference on Holography Applica- tions,1986: Peking, China. International Confer- ence on Holography Applications: 2-4 July, 1986, Beijing, China . Dahang Wang, chair. Jingtang Ke, Ryszard J. Pryputniewicz, eds. Sponsored by COS-Chinese Optical Society...Bellingham, WA: SPIE,1987. Series title: Proceedings ofSPIE- the International Society for Optical Engineering; v. 673.

International Conference on Holography, Optical Recording, and Processing of Information; 21-24 May 1989, Varna, Bulgaria; Y.N. Denisyuk, T.H. Jeong, chair/cochair; sponsored by The Bulgarian Academy of Sciences, ... let al.]. Bellingham, WA., USA : SPIE--the International Society for Optical Engineering, 1990.

International Congress on High-Speed Photogra- phy,11th:1974: Imperial College, London. High/speed photography: proceedings of the eleventh International Congress on High Speed Photogra · phy, Imperial College, University of London, Sep- tember 1974. P. J. Rolls, ed. London: Chapman & Hall: distributed in the USA by the Society of Photo-Optical Instrumentation Engineers, 1975.

International Congress on High Speed Photography,12th:1976: Toronto, Canada. Proceedings of the 12th International Congress on High Speed Photography (Photonics), Toronto, Can- ada, 1-7 August 1976. Martin C. Richardson. Bellingham, WA: Society of Photo-Optical Instrumentation Engineers, 1977. Series title: SPIEv.97.

International Congress on High Speed Photog- raphy and Photonics, 13th: 1978: Tokyo. Pr0- ceedings of the 13th International Congress on High Speed Photography and Photonics-Tokyo, 20-25 August 1978. Shin-ichi Hyodo, ed. Tokyo: Japan Society of Precision Engineering; [New York] : distributed (outaide Japan) by Society of Photo-Optical   Instru- mentation Engineers, 1979. Series title: SPIE v. 189.

International   Optical   Computing Confer- ence,1974: Zurich. Digest of papers. New York, Institute of Electrical and Electronics Engineers,1974.

International Optical Computing Confer- ence,1975: Washington, D.C. Digest of papers: International Optical Computing Conference, April 23-25, 1975, Washington, D.C. Sponsored by the Computer Society of the Institute of Electrical and Electronic Engineers, in coopera- tion...New York: Institute of Electrical and Electronics Engineers, 1975.

International Symposium on Acoustical Holog- raphy: Acoustical holography. New York: Ple- num Press, 1967.

International Symposium on Acoustical Holography.1st: 1967: Huntington Beach, CA. Acous-Laser Holography in Geophysics. Takemoto, Shuzo, Editor. New York, WileY,1989.
Lehman, Edward J. Applications of Hologra- phy: a bibliography with abstracts. Springfield, VA.: NTIS, 1975.

International Symposium on Acoustical Holog- raphy and Imaging,7th: 1976: Chicago, IL. Recent advances in ultrasonic visualization. Lawrence W. Kessler,ed. New York: Plenum Press, 1977. Series title: Interna- tional Sympo- sium on Acoustical Holography and Imaging Acoustical holography; v. 7.

International Symposium on Acoustical Holog- raphy and Imaging, 8th, Key Biscayne, FL, 1978. Ultrasonic visualization and character- ization. A. F. Metherell, ed. New York: Plenum Press,1980. Series title: International Sympo- sium on Acoustical Holography and Imaging Acoustical imaging; v. 8.

Klein, H. Arthur. Holography. With an introd. to the optics of diffraction, interference, and phase differences 1st ed. Philadelphia: Lippincott, 1970.Series title: Introducing modern science.

Kock, Winston E. Radar, sonar, and holography: an introduction. New York: Academic Press, 1973.

Kostelanetz, Richard. On Holography. RK Edi- tions,1979.

Kock, Winston E. Lasers and holographY; an intro- duction to coherent optics 1st ed.Garden City, N.Y., Doubleday, 1969.

Kurtz, Maurice K. Study of potential applica- tion of holographic techniques to mapping; final technical report. Lafayette, IN: Purdue Research Foundation, Purdue University, 1971. Lehmann, Matt. Holography; technique and practice. London: Focal Press, 1970. Series title: The Focal library.

Light and its uses: making and using lasers, holograms, interferometers,and instruments of disper- sion: readings from Scientific American. San Francisco: W. H. Freeman, 1980.

Lingenfelder, P. G. Holography manual; a com- pilation of laboratory techniques commonly used in the construction of holograms including refinements developed at NELC.. San Diego, CA: Naval Electronics Laboratory Center, 1969. Series title: NELC Techni- cal document 47.

McNair, Don. How to make holograms. 1st ed. Blue Ridge Sumnrit, PA: Tab Books,1983.
"A Multi-frequency synthetic detecting hologra- phy with high depth resolution". Peking, China: The Research Group of Holography, Chinese Academy of Geological Sciences, [s.n.), 1976.

NATO Advanced Study Institute on Optical and Acoustical Holography, 1971: Milan. Optical and acoustical holography; proceedings of the NATO Advanced Study Institute on Optical and Acoustical Holography, Milan, Italy, May 24- June 4, 1971. Ezio Camatini,ed. New York: Ple- num Press, 1972.

Okoshi, Takanori. Three-dimensional imaging tech- niques. New York: Academic Press, 1976.

Optical & Acoustical Holography. Camatini, E.- Edi- tor. New York: Plenum Publishing,1972.

Optical Computing Symposium, Darien, Conn., 1972. Digest of papers presented at the 1972 one-day-in- depth Optical Computing Sympo- sium, April 12, 1972 at the Noroton School, Darien, Connecticut. Naval Underwater Sys- tems Center and IEEE Computer Society, East- ern... [s.I.) Institute of Electrical Engineers, 1972.

Optical Information Processing and Hologra- phy, CatheY,W.Thomas.,Editor. Series Titile: Pure & Ap- plied Optics Series. New York: Wiley, 1974.

Optics and photonics applied to three-dimen- sional imagery (Image 3-0): presented as part of the Op- tics, Phototonics, and !conics Engi- neering Meet- ing (OPIEM), November 26-30, 1979, Strasbourg, France. Bellingham, WA: Society of Photo-Optical In- strumentation Engi- neers,1980. Series ti tie: Society of Photo-optical Instrumentation Engineers Seminar proceedings ; v. 212.

Optics in engineering measurement: 3-6 December 1985, Cannes, France. William F. Fagan, chair,ed. Organized by SPIE--the Inter- national Society for Optical Engineering, ANRT--Association Nationale de...Bellingham, WA: SPIE--the International So- ciety for Optical Engineering,1986. Series title: Pro- ceedings of SPIE--the International Society for Opti- cal Engineering; v. 599.

Optics in entertainment: January 20-21, 1983, Loa Angeles,   California. Chris   Outwater, chair,ed. Bellingham, WA: SPIE--the Interna- tional Society for Optical Engineering,1983. Series title: Proceed- ings of SPIE--the Interna- tional Society for Optical Engineering; v. 391.

Optics in entertainment: January 26-27, 1984, Loa Angeles, CA. Chris Outwater, chair, ed. Bellingham, WA: SPIE--the International Soci- ety for Optical Engineering,1984.Series title: Proceedings of SPIE-- the International Society for Optical Engineering; v. 462.

Optics, Photonics, and Iconics Engineering Meeting, 1979: Strasbourg, France. Optics and photonics ap- plied to three-dimensional imagery (IMAGE 3-0): presented as part of the Optics, Photonics, and Icon- ics Engineering Meeting (OPIEM), November 26-30, 1979, Strasbourg, France. Bellingham, WA: Society of Photo-optical Instrumentation Engineers,198O. Series title: Society of Photo-optical Instrumentation Engineers Proceedings; v. 212.

Optics Today. John N. Howard, ed. New York, N.Y: American Institute of Physics, 1986. Series title: Readings from Physics today; no. 3.

Ostrovskii, IU. I. Holography and its applica- tion. Translated from the Russian by G. Leib. Moscow: Mir, 1977.

Outwater, Chris. and Eric Van Hamersveld. Guide to practical holography. Beverly Hills, CA: Pentangle Press,1974.

Pattern recognition studies: seminar-in-depth, pro- ceedings / Society of Photo-Optical Instru- menta- tion Engineers. [Redondo Beach, Calif.l : the Society, [cl9691. Series title: S.P.I.E. semi- nar proceedings; v. 18.

Pattern Recognition & Acoustical Imaging. Fer- rari, Editor. Bellingham, WA: SPIE,1987.

Periodic structures, gratings, moire patterns, and diffraction phenomena: July 29-August 1, 1980, Sa~ Diego, CA. C.H. Chi, E.G. Loewen, C.L. O'Bryan III, eds. Bellingham, WA: Society of Photo-optical Instru- mentation Engi- neers,1981. Series title: Proceedings of the Soci- ety of Photo-optical Instrumentation En- gineers ; v. 240.

Pethick, J. On holography and a way to make holo- grams. Ontario: Belltower Enterprises, 1971.

Photonics applied to nuclear physics, 2: pro- ceedings; Strasbourg, Council of Europe, 5-7 December 1984. Geneva: CERN, 1985. Series titie: Nucleophot.

Photopolymer device physics, chemistry, and applica- tions: 17- 19 January 1990, Los Angeles, California. Roger A. Lessard, chair/editor; spon- sored by SPIE- -the International Society for Optical Engineering. Bellingham, WA., USA: SPIE--the International So- ciety for Optical Engineering, 1990.

Photorefractive materials and their applica- tions. P. Gunter, J.-P Huignard, eds. Contribu- tions by A.M. Glass... let a1.1. Berlin: Springer- Verlag, 1988. Se- ries title: Topics in applied physics; v. 61, etc.
tical holography; proceedings. Plenum Press, 1969. New York:

Pietsch. Paul. ShuffHebrain. Boston: Houghton Mifflin. 1981.

Pisa. Edward J.• S. Spinak & A.F.Metherell. Color acoustical holography. Huntington Beach. CA: Douglas Advanced Research Laboratories. 1969. Series title: Douglas Advanced Research Laboratories. Research communication 109.

Practical holography: 21-22 January 1986. Los Angeles. CA. Tung H. Jeong. Jacques E. Ludman chair. eds. Presented in cooperation with Ameri- can Association of Physicists in Medicine... ret al.). Bellingham. WA: SPIE-The International Society for Optical Engineering. 1986. Series title: Proceedings of SPIE--the International Society for Optical Engineering; v. 615.

Practical holography II: 13-14 January 1987. Los Angeles. CA. Tung H. Jeong. chair/ed. Sponsored by SPIE--the International Society for Optical Engineering. in cooperation with Center for.. Bellingham.W A: SPIE--the International Society for Optical Engineering.1987. Series title: Pr0- ceedings of SPIE--the International Society for Optical Engineering; v. 747.

Practical holography III : 17-18 January 1989. Los Angeles. CA. Stephen A. Benton. chair/editor ; sponsored by SPIE--the International Society for Optical Engineering ; cooperating organiza- tions. Applied.. Bellingham. WA.• USA: SPIE. 1989. Series title: Proceedings ofSPIE--the Inter- national Society for Optical Engineering; v. 1051.

Proceedings of the information processing and holography symposium ICALEO 83. Symposium heads: David Casasent. Milton T. Chang. Orga- nized with American Society of Metals... ret al.). Sponsored by Laser Institute... Toledo. OH: The Institute. 1984. Series title: LIA (Series); v. 41.

Proceedings of the Inspection. Measunnent [sic] and Control and Laser Diagnostics and Photo- chemistry. ICALEO '84. Donald Sweeney, Robert LUcht. eds. Organized in cooperation with... The American... Toledo, OH: LIA-Laser Institute of America. 1985. Series title: LIA (Series) ; v. 45. 47.

Processing and display of three-dimensional data: August 26-27. 1982, San Diego. CA. James J. Pearson, chair,ed. Bellingham. WA: SPIE--The International Society for Optical Engineer- ing,1983. Series title:Proceedings of SPIE--the International Society for Optical Engineering; v. 367.

Processing and display of three-dimensional data II : August 23-24. 1984. San Diego, CA. James J . Pearson. chair.ed. Cooperating organizations, Optical Sciences Center. University of Arizona•Bellingham, WA: SPIE--the International Society for Optical Engineering.1984. Series title: Pr0- ceedings of SPIE--the International Society for Optical Engineering; v. 507.

Progress in holographic applications: 5-6 Decem- ber 1985. Cannes. France. Jean Ebbeni. chair, ed. Organized by SPIE--the International Society for Optical Engineering. ANRT--Association Nation- ale de la... Bellingham. WA: SPIE--the Interna- tional Society for Optical Engineering,1986. Series title: Proceed- ings of SPIE--the Interna- tional Society for Optical Engineering; v. 600.

Progress in holography: 31 March-2 April 1987. The Hague. The Netherlands. Jean Ebbeni. chair/ed. Or- ganized by ANRT--Association natio- nale de la re- cherche technique, SPIE--The Inter- national Society for...Bellingham,W A: SPIE. 1987. Series title: Proceedings of SPIE--The International Society for Optical Engineering; v. 812.

"Project Search". Subcomrrrittee on Feasibility of Au- tomated Fingerprint IdentificationlVerification. An experiment to determine the feasibility of holograph assistance to fingerprint identification. Sacramento. CA: 1972. Series title: Project Search Technical re- port, no. 6.

Recent advances in holography III: February 4-5, 1980. Los Angeles, CA. Tzuo-Chang Lee. Poohsan N. Tamura, cds. Bellingham, WA: Society of Photo- optical Instrumentation Engineers. 1980. Series title: Society of Photo-optical Instrumenta- tion Engineers Proceedings; v. 215.

Saxby. Graham. Practical holography. New York. N.Y.: Prentice-Hall International, 1988.

_ _ _ . Holograms: How to Make & Use Them. Mas- son, France: Focal Press. 1980.

Saxby. John. Holograms. New York: Focal Press. 1980.

Schultz, Jerold M. Diffraction for Materials Sci- en- tists. New York: Prentice-Hall, 1982.

Schumann. Walter and J . P. Zurcher, D.Cuche. Ho- lography and deformation analysis. Berlin: Springer- V erlag. 1985. Series Series in Optical Sci- ences; v. 46.

Smith, Howard Michael. Principles of hologra-phy. New York, Wiley-Interscience,1969.

_ _ _ . Principles of holography. 2d ed. New York: WileY,1975.

Solem, Johndale C. High-intensity X-ray holography: an approach to high·resolution snapshot imaging of biological specimens. Los Alamos. N.M.: Los Alamos National Laboratory.1982.

Solymar. L. and D.J . Cooke. Volume holography and volume gratings. London: Academic Press, 1981.

Soroko. Lev Markovich. Holography and coher- ent optics. Translated from Russian by Albin Tybulewicz; with a foreword by George W. Stroke. New York: Ple- num Press. 1980.

Sources of Physics Teaching: Atomic Energy. Holog- raphy, Electrostatics; Vol. 4. Noakes, G. R.- Editor. New York: Coronet Books.1970.

Spencer John R. Holographic Infonnation Stor- age and Retrieval. England: National Repro- graphic. 1975.

Stroke, George W. An introduction to coherent optics and holography. New York: Academic Press. 1966.

_ _ _ . An introduction to coherent optics and hologra- phy. 2d ed. New York: Academic Press. 1969.

Symposium on Applications of Holography in Me- chanics.1971: University of Southern Califor- nia. Symposium on Applications of Holography in Me- chanics. W. G. Gottenberg. ed. New York: American Society of Mechanical Engineers, 1971.

Three-dimensional imaging; April 21-22. 1983, Ge- neva, Switzerland. Jean Ebbeni. Andre Mon- fils. chainnen-editorB . Bellingham. WA: SPIE-- the In- ternational Society for Optical Engineer- ing.1983. Series title: Proceedings of SPIE--the International Society for Optical Engineering ; v. 402.

Three-dimensional imaging: August 25-26. 1977. San Diego. CA. Stephen A. Benton. ed. Presented by the Society of Photo-optical Instrumentation Engineers. in conjunction with the IEEE Computer... Belling- ham. WA: SPIE. 1977. Series title: Society of Photo- optical Instrumentation Engi- neers Proceedings; v. 20.

Ultrasonic Imaging & Holography: Medical. So- nar. & Optical Applications. Stroke. George W.,and Jumpei Tsujiuchi. Editors. New York: Ple- num Publishing.1974.

United States-Japan Science Cooperation Seminar on Pattern Infonnation Processing in Ultra- sonic Imaging. 3rd: 1973: University of Hawaii. Ultrasonic imaging and holography: medical. sonar, and optical applications: [proceedings]. George W. Stroke... ret al.),ed. New York: Plenum Press, 1974.

United States-Japan Seminar on Infonnation Pro- cessing by Holography, 2nd :1969 : Washing- ton, D.C. Applications of holography; proceed- ings. Euval S. Barrekette. ed. New York: Plenum Press, 1971.

Unterseher, Fred. Jeannene Hansen and Bob Schlesinger. Holography handbook: making holo- grams the easy way. Berkeley. CA: Ross Books. 1982.

Vasilenko, G. 1. (Georgii Ivanovich) and L.M. Tsibul'kin. Image recognition by holography. Trans- lated from Russian by Albin Tybulewicz. New York : Plenum Press/Consultants Bureau. 1989.

Weinstein. L. Albertovich. Theory of Diffraction & the Factorization Method: Generalized Wiener- Hopf Technique.Golem Publications,1969. Elec- tromag- netics Series. Vol. 3.

Wenyon. Michael. Understanding holography. New York: Arco Pub. Co.• 1978.

_ _ _ . Understanding holography. Newton Abbot, Eng.: David & Charles.1978.

_ _ _ . Understanding holography. 2nd Areo ed. New York: Areo Pub.• 1985.

Wolff J . (et alia). Light Fantastic; Lasers and Holography Explained. England: Gordon Fraser/ Bergstrom & Boyle. 1977

Yaroslavskii. L. P.• and N. S. Merzlyakov. Meth- ods of Digital Holography. New York: Plenum Publish- ing, 1980.

Yu, Francis T.S. Introduction to diffraction. infor- mation processing, and holography. Cambridge, MA: MIT Press • 1973.

General Holography: Information—dissertations

Coello-Vera. Agustin Elias. "Scanned acoustic imag- ing in the ocean: a study of holographic-like systems and their limi tations". 1978.

Elliott, John Douglas. "Computer simulation of the holographic image degradation due to transmission of the signal through a random noise media".1971.

Eu, James Kim-Tzong. "Studies in spatial filtering". 1974.

Fischer, Wolfgang Klaus. Methods for acoustic holography and acoustic measurements. Newark, N.J. : [s .n.l, 1972.

George, Daweel Joseph. "Holography as applied to jet breakup and an analytical method for reducing holographic droplet data". 1972.

Kurtz, Maurice K. "Potential uses ofholography in photogrammetric mapping". 1971.

Landry, Caliste John. "Ultrasonic imaging by Brillouin-Bragg diflioaction : development of an operational system with prospective applications in medical diagnosis and material testing". 1972.

Lee, Hua. "Development and analysis of the back-projection method for acoustical imaging". 1980.

Liu, Charles Yau-ehi. "Some topics in holographic image formation". 1974.

Mensa, Dean L."Techniques for microwave imaging" .1980.

Powers, John Patrick. "Some aspects of the application of Bragg diflTaction of laser light to the imaging and probing of acoustic fields". 1970.

Ramos, George Urban. "1. On the fast fourier transform; II. On the computations in digital holography". 1970.

Schlussler, Larry. "Improvement of the horizontal resolution of a Bragg-diflTaction imaging system and motion limitations of a holographic system". 1978.

Schueler, Carl Frederick. "Development and applications of computer-assisted acoustic holography". 1980.

Schwank, James Ralph. "Refractive holography". 1974.

Sherman, George Charles. "Wavefront reconstruction and ita application to the study of the optical properties of atmospheric aerosols". 1969.

Shuman, Curtis Alan. "Holographic imaging through moving diffusive media". 1973.

___ . "Holographic imaging through moving scatters".1972.

Stone, William Ross. 'The concept, design, and operation of a demonstration holographic radio camera". 1978.

Strand, Timothy C. "Comparison of analog and binary holographic data storage". 1973.

Sutton, Jerry Lee. "Broadband acoustic imaging". 1974.

Tricoles, Gus Peter. "Some topics in microwave holography". 1971.

Tse, Nie But. "Digital reconstruction of acoustic holograms". 1979.

Vourgourakis, Emmanuel John's. "Coherence limitations on holographic systems". 1967.

Wang, Keith Yu-Chih. 'Threshold contrasta for various acoustic imaging systems". 1972.

Wollman, Michael Thomas. "An experimental acoustical holographic system for eventual use in the ocean". 1975.

Industrial Holography

Acoustical Holography; Vol. 1. Metherell, A.F., Editor. New York: Plenum Publishing,I969.

Acoustical Holography; Vol. 2. Metherell, A.F., Editor. New York: Plenum Publishing, 1970.

Acoustical Holography; Vol. 3. Metherell, A.r!F., Editor. New York: Plenum Publishing,1971.

Acoustical Holography; Vol. 4. Wade, Glen, Editor. New York: Plenum Publishing,1972.

Acoustical Holography; Vol. 5. Green, Philip S.,Editor. Plenum Publishing,1974.

Acoustical Holography; Vol. 6. Booth, N., Editor. New York: Plenum Publishing,1975.

Acoustical Holography; Vol. 7. Kessler, L.W., Editor. New York: Plenum Publishing, 1977.

Acoustical Imaging; Vol. 8. Metherell, A.F., Editor. New York: Plenum Publishing,I980.

Acoustical Imaging; Vol. 9. Wang, Keith, Editor. New York: Plenum Publishing, 1980.

Acoustical Imaging; Vol. 11. Powers, John P.,Editor. New York: Plenum Publishing, 1982.

Acoustical Imaging; Vol. 11. Ash, Eric A. and C.R. Hill ,Editors. New York: Plenum Publishing, 1983.

Acoustical Imaging, Vol. 15. Jones, Hugh W., Editor. Plenum Publishing,1987.

Acoustic Imaging: cameras, microscopes, phased arrays, and holographic systems. Glen Wade, ed. New York: Plenum Press, 1976.

Acoustic Surface Wave & Acousto-Optic Devices. Kallard, Thomas-Editor. Series Title: State of the Art Review Series; Vol. 4. New York: Optosonic Press,1971.

Aldridge, Edward E. Acoustical holography. Watford: Merrow Publishing Co. Ltd., 1971. Series title: Merrow monographs, practical science series 1.

Applications of Holography. Barrekette, E. S., Editor. New York: Plenum Publishing,1971.

Applications of Holography in Mechanics: Symposium, University of Southern California, 1971. Symposium Staff; Gottenberg, W. G., Editors. Books on Demand UMI, Reprint of 1971 edition.

Applications of Holography & Optical Data Processing: Proceedings of an International Conference, Jerusalem, 1976. Marom, E.; Avnear Wiener and A.A. Friesem, Editors. London: Pergamon Press,1977.

Applications oflasers to photography and information handling; proceedings, two-day seminar.

Richard D. Murray, ed.Washington, DC: Society of Photographic Scientista and Engineers, 1968.

Basov, N. G. Lasers & Holographic Data Processing. USSR: Mir Publications, 1985.

___ . Lasers & Holographic Data Processing. England, Colletts: State Mutual Books,1984.

Beiser, Leo. Holographic scanning. New York: WileY,1988. Series title: Wiley series in pure and applied optics.

Brcic, Vlatko. Application of holography and hologram interferometry to photoelasticity: lectures held at the Department for Mechanics of Deformable Bodies. 2d ed. Wien: Springer-Verlag, 1974. Series title: Courses and lectures; no.7.

Bristol University Electron Microscopy Group. Convergent Beam Electron DiflTaction of Alloy Phases. Mansfield, J., Editor. A Hilger UK: Taylor & Francis, 1984.

Business Communications Staff. Holography: New Commercial Opportunities. [s.l]: BCC,1986.

Butters, John N. Holography and ita technology. London: P. Peregrinus,1971; Published on behalf of the Institution of Electrical Engineers. Series title: Institution of Electrical Engineers I.E.E. monograph, series, 8.

Ceccon, Harry L. Holographic techniques for nondestructive testing of tires. Washington, D.C: National Highway Traffic Safety Administration, 1972.

Conferer_ce on Holographic Instrumentation Applications,1970: Ames Research Center. Holographic instrumentation applications. Prepared by NASA Ames Research Center. Boris Ragent and Richard M. Brown, eds_ Washington, DC: Scientific and Technical Information Division, National Aeronautics and Space Administration,1970. Series title: NASA SP; 248.

Dirtoft, Ingegard. Holography: A New Method for Deformation Analysis of Upper Complete Dentures in Vitro & in Vivo. New York: Coronet Books,I985.

An External Interface for Processing 3-D Holographic & X-Ray Images. Juptner, W., Editor New York: Springer-VerlaglResearch Reports,1989.

Flow visualization and aero-optics in simulated environments:21-22 May 1987, Orlando, Florida. H. Thomas Bentley III, chair/ed. Sponsored by SPIE--the International Society for Optical Engineering. Bellingham, WA: SPIE--the International Society for Optical Engineering,1987. Series title: Proceedings of SPIE--the International Society for Optical Engineering; v. 788.

Hartman, W. F. Acoustic Emission: Advances in Acoustic Emission. American Society for Nondestructive Testing, 1981.

Holographic data nondestructive testing : October 4-8, 1982, Croatia Hotel de Luxe, Dubrovnik, Yugoslavia. Dalibor Vukicevic, chair,ed. Sponsored by the International Commission for Optics (ICO) [andl ... Bellingham,

WA: SPIE--the International Society for Optical Engineering,1983. Series title: Proceedings of SPIE--the International Society for Optical Engineering; v. 370.

Holographic nondestructive testing. Robert K. Erf, ed. New York: Academic Press, 1974.

Holographic nondestructive testing: status and comparison with conventional methods: 23-24 January 1986, Los Angeles, California. Charles M. Vest, chair, ed. Presented in cooperation with American Association...Bellingham, WA: SPIE-- the International Society for Optical Engineer- ing, 1986. Series title: Proceedings of SPIE--the International Society for Optical Engineering; v. 604.Series title: SPIE critical reviews of technol- ogy series; 15th.

Holographic Nondestructive Testing. Erf, Robert K., Editor. New York: Academic Press,1974.

Holographic Nondestructive Testing: Critical Review of Technology. Vest, Charles., Editor. Bellingham, WA: SPIE,1986.

Holography in Medicine: International Sympo- sium Proceedings. Greguss, Pal. Editor.I.P.C.- Sci.& Technology, 1976.

Holography in Medicine & Biology. Von Bally, G., Editor. New York: Springer-Verlag,1979. Series Title: Springer Series in Optical Sciences,; Vol. 18.

Industrial applications of holographic nonde- structive testing: May 3-5, 1982, Brussels. J. Ebbeni, chair, ed. Sponsored by SPIE- the Inter- national Society for Optical Engineering; with the support... Bellingham, WA: SPIE--the Inter- national Society for Optical Engineering,1982. Series title: Proceed- ings of SPIE--the Interna- tional Society for Optical Engineering; v. 349.

Industrial Applications of Holography. Robillard, Jean, Editor. Oxford, UK: Oxford University Press, 1989 and 1990.

Industrial Radiography Holography. American Society for Nondestructive Testing, 1983.

International Symposium on Holography in Biomedical Sciences,1973: New York. Holography in medicine: proceedings of the International Sym- posium on Holography in Biomedical Sciences, New York, 1973. PAl Greguss, ed. Guildford, Eng: IPC Science and Technology Press, 1975.

International Workshop on Holography in Medi- cine and Biology,1979: Munster, Germany. Holog- raphy in medicine and biology : proceedings of the International Workshop, Munster, Fed. Rep. of Germany, March 14-15, 1979. G. von Bally, ed. Berlin: Spring- er-V erlag, 1979. Series title: Springer Se- ries in Optical Sciences; v 18.

Jones, Robert; and Catherine M. Wykes. Holo- graphic and Speckle Interferometry. Cambridge: Cam- bridge University Press,l983.

Jones, Robert; and Catherine M. Wykes. Holo- graphic & Speckle Interferometry. Cambridge University Press,1989. Series title: Cambridge Studies in Modern Optics,; No.6.

Nondestructive holographic techniques for struc- tures inspection. R. K. Erf...[et al.). Wright- Patterson Air Force Base, OH: Air Force Materi- als Laboratory, Air Force Systems Command, 1972.

Oetrovsky, Y. I. and M.M. Butusov. Interferome- try by Holography. New York: Springer-Verlag, 1980. Series Title: Springer Series in Optical Sci- ences,; Vol. 20.

Schumann, Walter, and M. Dubas. Holographic In- terferometry: From the Scope of Deformation Analy- sis of Opaque Bodies. Series Title: Springer Series in Optical Sciences,; Vol. 16. New York: Springer- Verlag,1979.

Spanner, Jack C. Acoustic Emission Testing: Acous- tic Emission: Techniques & Applications. American Society for Nondestructive Test- ing,1974.

Thpical Meeting on Hologram Interferometry and Speckle Metrology,1980, June 2-4 : North Fal- mouth, MA. A digest of technical papers pre- sented at the Thpical Meeting on Hologram Interferometry and Speckle Metrology, June 2-4, 1980, Sea Crest Hotel, North Falmouth, Cape Cod, MA. [s.I.): Optical Society of America,1980.

Vest, Charles M. Holographic interferometry. New York: Wiley, 1979. Series title: Wiley series in pure and applied optics.

Industrial Holography Dissertations

Dallas, William John. "Computer holograms: improv- ing the breed". 1971.

Dzekov, Thmislav Angel. "Microwave holographic imaging of aircraft with spacebome illuminating source". 1976.

Matthews, Barbara Kubitz. "Application of holo- graphic methods to the analysis of flexural vibra- tions ofannular sector plates". 1976.

Su, Kung-Yen. "I'he fabrication of an opto-acous- tic transducer for real-time diagnostic imaging sys- tems", 1982.

Holography Marketing

Holography, 1971-72. Kallard, Thomas-Editor. Se- ries Title: State of the Art Review Series, Vol. 5. New York: Optosonic Press,1972.

Holography: Exploiting the Leading-Edge Devel- op- ments. Chicago: Technical Insights, 1987.

Holography Marketplace. Ross, Franz and Eliza- beth Yerkes, Eds. Berkeley , CA: Ross Books, 1989.

Holography Marketplace, 2nd Edition. Ross, Franz and Elizabeth Yerkes, Eds. Berkeley, CA: Ross Books,1990.

Holography Marketa. International Resource Devel- opmen ts,1984.

Industrial and commercial applications of holog- raphy: August 24-25,1982, San Diego, CA. Milton Chang, chair,editor. Bellingham, WA: SPIE--the In- ternational Society for Optical Engineer- ing,1983. Series title: Proceedings of SPIE--the International Society for Optical Engineering ; v. 353.

Kallard, Thomas. Holography; state of the art review, 1969. New York: Optosonic Press,1969. Series title: State of the art review, 1.

___. Holography; state of the art review ... 1970. Ho- lography in1970: an overview by Dr. Dennis Gabor. New York: Optosonic Press,1970. Series title: State of the art review, no. 3.

Miller, Richard K. (et alia).Holography. New York: Future Tech Surveys, 1989. Series title: A Survey on Technology & Markets Ser.,; No. 51

Periodicals

Acoustical Holography. International Symposium on Acoustical Holography. New York: Plenum Press. v.1-7, (1969-1977).

Acoustical holography, proceedings. Interna· tional Symposium on Acoustical Holography and Imaging. (-1973).

Acoustical imaging and holography. New York: Crane, Russak,1978-1979.

Advances in holography. New York, M. Dekker, 1975-76.

Acoustical holography; [proceedings): Acoustical im- aging, 1978. International Symposium on Acoustical Holography and Imaging. New York: Plenum Press, 1978.

Afterimage. Rochester, NY: Visual Studies Work- shop, V.12, no.7, Feb. 1985. See: A. Sargent- Wooster, "Manhattan shortcuts, Hamet Casdin- Silver's 'Thresholds'" p 19.

Fundamentals and applications of optical data pro- cessing and holography. Ann Arbor: Univer- sity Michigan Engineering Summer Conferences.

Holography News. (newsletter). Washington, DC.: Louis Kontnick, since 1987. (see listing for address).

The Holo-gram. (newsletter). Allentown, PA: Frank DeFreitas, since 1983. (See listing for address).

New Scientist. London, England: 1977. See: R. Weale "Art: Holography by H.Casdin-Silver" (June 29).

REFERENCES IN LANGUAGES OTHER THAN ENGLISH

French Titles ~

Caussignac, Jean Marie. Visualisation d'ecoule- ments aerodynamiques dans les compresseurs par interferometrie holographique. Chatillon, France: Of- fice national d'etudes et de recherches aerospatiales, 1972. Series title: France. Office national d'etudes et de recherches aerospatiales Note technique, 190.

Francon, M. Holographie. Paris: Masson, 1969. Se- ries title: Recherche appliquee.

International Symposium of Holography, Besan- con, France, 1970. Applications de I'holographie; comptes rendus du Symposium international d'holographie. Applications of holography; pro- ceedings of the In- ternational Symposium of Holography. Besancon 6-11 juillet 1970. Besan- con: Laboratoire de physique generale et optique, Universite de Besancon,1970.

Pinson. G.. A. Demailly, andD. Favre.La Pensee: approche Ibolographique. Lyon: Presses universitaires de JLyon. 1985. Series title: Col- lection Science des systemes. Serie theorie des systemes.

Voropaiev. N. Dictionnaire d'Electronique Quantique. Holographie et Optoelectronique. France: French & European.l983

German Titles

Claus. Jurgen. ChippppKunst: Computer. Holographie. Kybernetik. Laser. Originalausg. FrankfurVM.: U11stein.1985. Series title: U11- stein Materilien.

Kiemle. Horst. [undl Dieter Ross. Einfuhru.ng in die Technik der Holographie. Frankfurt am Main. Akademische V erlagsanstalt. 1969. Series title: Technisch-physikalische Sammlung Bd.8.

Laserbeugung an elektronenmikroskopischen Aufnahmen. Ludwig Reimer ... [et al.l. Opladen: Westdeutscher Verlag. 1973. Series title: Forschungsberichte des Landes Nor- drhein-Westfalen; Nr. 2314.

Licht-Blicke : Holographie. die 3. Dimension fur Technik und kunst : [Ausstellungl 7. Juni-30. September 1984. Deutsches Filmmuseum Frankfurt am Main. Schirmherr. Bundesprasi- dent a. D. Walter Scheel;...Frankfurt am Maain: [Deutsches Filmmuseuml.1984. Interviews with: S.Benton. M.Benyon. R.Berkhout. H.Cas- din-Silver. F.Mazzero. S.Moree. N.Phillips. G. Schneider-Siemssen. D.Schweitzer. Articles by: M. Schneckenburger. et alia.

Mehr Licht: Kunstlerhologramme und Lichtob- jekte = More light : [artists's [sicl holograms and light objectsl. herausgegeben von Achim Lipp und Peter Zec. Hamburg: Fielmann im E. Kabel Verlag. 1985.

Menzel. Eric. W. Mirande [undl I.Weingartner. Fourier-Optik und Holographie Wien: Springer-Verlag. 1973.

Optoelektronik in der Technik : Vortrage des 6. Internationalen Kongresses Laser 83 Optoelek- tronik = Optoelectronics in engineering : pro- reedings of the 6th International Congress. Laser 83 Optoelektronik I herausgegeben. Ber- lin: Springer-Verlag. 1984.

Schreier. Dietmar unter Mitarbeit von W. Hase..[et al.l Synthetische Holografie. Wein- heim: Physik-Verlag. 1984.

Universitatsbibliothek Jena. Zusammenstel- lung in- und auslandischer Patentschriften auf dem Gebiet der Holographie. Berichtszeit: 1948-1970. Gesamtleitung: Konrad Marwinski. Informationsabt.• 1971. Series title: Universi- tatsbibliothek Jena Bibliographische Mitteilun- gen, Nr. 12.

Voropaev. N. D. Woerterbuch der Quantenelek- tronik. Holographie und Optoelektronik French & European.1983

Zec. Peter. Holographie: Geschichte. Technik. Kunst. Koln: DuMont.1987.

Portuguese (Articles)

Catálogo of VII Salão Nacional de Artes Plasti- CBS. "As tn\B dimensOes do signo vemal". Edu- ardo Kac. Museu de Arte Moderna. Rio de Janeiro. pp. 43-44. 1984.

Folha de SAo Paulo. "Na holografia. 0 cinema tridimensional do futuro". Eduardo Kac. SAo Paulo. August 21. 1985.

Folha de SAo Paulo. "Harriet. pioneira na arte do laser. expOO na Bienal". Eduardo Kac. SAo Paulo. October 2. 1985.

Folha de SAo Paulo. "Holografia impressa sera prduzida comercialmente". Eduardo Kac. SAo Paulo. May 20. 1987.

Folha de SAo Paulo. "Ingl~ m08tra seu per- curso na holografia". Marion Strecker. SAo Paulo. June 20. 1989.

Jornal do Brasil. "A Arte da Sintese n08 holopoemas". Reynaldo Roels Jr.• Rio de Janeiro, Septemebr 29, 1985.

Jornal do Brasil. "A holografia da urn pasBO II frente". Eduardo Kac, Rio de Janeiro, Septem- ber 29.1985.

o Estado de SAo Paulo. "Cariocas inovam holografia". Sergio Adeodao. SAo Paulo. Novem- ber 27,1988.

o Globo. "Primeira m08tra de arte high tech". Frederico de Morais, Rio de Janeiro, April 6, 1986.

o Globo. "Holografia: 0 ator sai da tela e senta perto do espectador". Sheila Kaplan. Rio de Janeiro. December 13. 1984.

o Globo. "0 Sonho holografico de Alexander". Eduardo Kac. Rio de Janeiro. November 11. 1987.

o Globo. "Holofractal, a arte no futuro". Ligia Canongia.Rio de Janeiro. November 22. 1988.

Russian Titles

Bakhrakh, L.D. i S.D. Kremenetskii. Metody izmerenii parametrov iizluchaiushchikh sistem v blizhnei zone. Leningrad : Izd-vo "Nauka", Leningradskoe otdelenie. 1985.

Bakhrakh. L.D i V.A. Makeeva. Primenenie golografii v meditsine i biologii. Leningrad : Nauka, 1977.

Barachevskii, V.A.Svoistva svetochuvstvitel'- nykh materialov i ikh primenenie v golografii: sbornik nauchnykh tru.dov. Otvetstvennyi redaktor Leningrad: Izd-vo "Nauka," Lenin- gradskoe otd-nie. 1987.

__ . Neserebrianye i neobychnye sredy dlia golografii. Leningrad: "Nauka". Leningradskoe otd-ie. 1978. "Inteli~ncia"
Denisiuk. IU.N. Opticheskaia golografiia : prakticheskie primeneniia. . Leningrad : Izd-vo "Nauka," Leningradskoe otd-nie. 1985.

_ _ _ ..Opticheskaia golografiia: [Sb. statei]. AN SSSR. Fiz. tekhn. in-t im. A.F. lotTe. Nauch. BOvet po probl. "Golografiia". Leningrad: Nauka. Leningr. otd-nie. 1979.

Derkach. M.F.Dinamicheskie spektry rechevykh signalov. L'vov: Izd-vo pri L'vovskom Gas. univer- sitete Izdatel'skogo ob"edineniia "Vyshcha shkola". 1983.

Fizicheskie osnovy i prikladnye voprosy golografii : tematicheskii sbornik: [materialy XVI Vsesoiuznoi shkoly po fizicheskim 08novam golografiV redaktory G.V. Skrotskii. B.G. Turukhano. N. Turukhanol. Leningrad: Aka- demiia nauk SSSR. Fiziko-tekhn. ins-t im. A.F. lotTe. 1984.

Gurevicha, S.B. i V.K Sokolova. Primenenie metodov opticheskoi obrabotki informatsii i golografii. Leningrad: LIIAF. 1980.

Gurevicha, S.B. Primenenie golografii v matsii. held in Riga, May 1980. Akademiia nauk SSSR. Ordena Lenina Fiziko-tekhnicheskii institut im. A.F. lotTe.

Gurevicha.S.B.. O.A . Potapova."Golografiia i opticheskaia obrabotka informatsii v geologii". Dokl. seminara. Leningrad: Akademiia Nauk SSSR. Fiziko-tekhnicheskii in-t im. A.F. lotTe. 1980.

IAkovkin. I. B. Difraktsia sveta na akus- ticheskikh poverkhnostnykh volnakh. otv. redaktor S.V. Bogdanov. Novosibirsk: Izd-vo "Nauka". Sibirskoe otd-ie. 1979.

IAroslavskii. L. P. and N.S. Merzliakov. TSi- frovaia golografiia. M08kva: Izd-vo "Nauka". 1982.

International School on Coherent Optics and Holography.2nd: 1981: Varna. Bulgaria. Integral'naia optika. volokonnaia optika i golografiia: materialy vtoroi Mezhdunarodnoi shkoly po kogerentnoi optike i golografii--Varna '81,28.09-03.10.1981. Varna, Bolgariia. Redakt- sionnaia kollegiia P. Simova.... Sofiia: Izd-vo Bolgarskoi akademii nauk. 1982.

Kirillov, N. I.Vysokorazreshaiushchie fotomate- rialy dlia golografii i protsessy ikh obrabotki . Moskva: Nauka, 1979.

Klimenko, I. S. Golografiia sfokusirovannykh izobrazhenii i spekl-interferometriia. M08kva: "Nauka," Glav. red. fiziko-matematicheskoi lit- rio 1985.

Klimkin. V. F. (Viktor Fedorovich) Opticheskie metody registratsii bystroprotekaiushchikh protsess- ov I V.F.

Klimkin, A.N. Papyrin, R.l. Soloukhin ; otvet· stvennyi redaktor N.G. Preobrazhenskii. Novosibirsk: Izd- vo "Nauka," Sibirskoe otd-nie, 1980.

Kulakov. Sergei Viktorovich. Akustoopticheskie us- troistva spektral'nogo i korreliatsionnogo analiza signalov. Akademiia nauk SSSR, Nauchnyi sovet po probleme"Golografiia", Fiz- iko-tekhnicheskii institut imeni A.F. lotTe. Len· ingrad :"Nauka", Leningrads- koe otd-nie, 1978.

Jornal do Brasil . Reynaldo Roels Jr.• Rio de Janeiro. December 1. 1986.

Jornal do Brasil. 'Um unic6rnio na matematica". Reynaldo Roels Jr.• Rio de Jan- eiro. November 29. 1988.

Jornal do Brasil. "Holografia gerada por com- putador". Marcelo Tognozzi. Rio de Janeiro. July 2. 1989.

Petrashen, G. I. Prodolzhenie volnovykh polei v zadachakh s seismorazvedki. Leningrad: "Nauka," Leningr. otd·nie, 1973.

Radiogolografiia i opticheskaia obrabotka infor· matsii v mikrovolnovoi tekhnike: [Sbornik statei). Akademiia nank SSSR, Otdelenie obsh· chei fiziki i astronomii, Nauchnyi sovet po prob- Ierne "Golografiia" Leningrad "Nauka," Leningradskoe otd-nie, 1980.

Soboleva, G.A. Registriruiushchie sredy dlia izo· brazitelnoi golografii i kinogolografii: [Sb. statei). AN SSSR, Otd-nie obshch. fiziki i astronomii, Nauch. sovet po probl. Golografiia. Leningrad: Nauka, Leningr. otd-nie, 1979.

Sokolov, A. V. i IA.A. Al'tmana. Primenenie metodov opticheskoi golografii dlia issledovaniia biologicheskikh mikroob"ektov. Leningrad: Nauka, Leningradskoe otd-nie,1978. Series title: Metody fiziologicheskikh issledovanii.

Voropaev, N. D. Anglo-ruaskii slovar' po kvanto- voi elektronike i golografii: Okolo 18000 termi- nov. Pod red. A. M. Leontovicha. Moskva: Rus.iaz., 1977.

Vsesoiuznaia shkola po golografii, 6th: 1974: Yerevan, Armenian S.S.R. Materialy VI Vsesoiuznoi shkoly po golografii:11-17 fevralia 1974 g. redaktory, G.V. Skrotskii, B.G. 'furukhano, N. Turukhano). Leningrad: LIIAF, 1974.

Vsesoiuznaia shkola po golografii, 7th: 1975: Rostov, R.S.F.S.R. Materialy VII Vsesoiuznoi shkoly po golografii: ianvar' 1975 g. [podgotovleny k pechatki N. Turukhano). Leningrad: Leningrad- skii in-t iadernoi fiziki, 1975.

Vsesoiuznaia shkola po fizicheskim osnovam golografii (14th:I982 : Dolgoprudnyi, R.S.F.S.R.). Prikladnye voproey golografii: tematicheskii sbornik. [redaktory, G.Y. Skrotskii, B.G. Turukh- ano, N. Turukhano). Leningrad: LIIAF, 1982.

Zel'dovich, B. lA, N .F. Pilipetskii, & V.V.Shkunov. Obrashchenie volnovogo fronta. Moskva: "Nauka," Glav. red. fiziko-matematicheskoi lit-ry, 1985.

Swedish Titles

Holgrafi: det 3-dimensionella mediet. New York: Museum of Holography; Stockholm: distribution, AVC, 1976.

Lasers And Holography

Advances in Laser Engineering: August 25-26. 1977, San Diego, California. Malcolm L. Stitch, Eric J . Woodbury, eds; presented by the Society of Photo-optical Instrumentation Engineers in con- junction with the IEEE Computer Society Inter- national Optical Computing Conference '77. Bellingham, WA: The Society, 1977.

Advances in nonlinear polymers and inorganic crystals, liquid crystals, and laser media: 20-21 August 1987, San Diego, California. Solomon Musikant, chair/editor; sponsored by SPIE-The International Society for Optical Engineering. Bellingham, WA, USA: SPIE--the International Society for Optical Engineering,1988.

Beck, Rasmus, W. Englisch, and K Gura. Table of Laser Lines in Gases and Vapors. 3d rev. and en!. ed. Berlin: Springer-Verlag, 1980. Series Title:

Springer Series in Optical Sciences; v. 2.

Bennett, William Ralph Jr. Atomic gas laser transition data: a critical evaluation. New York: IFIlPlenum,1979. Series: IFI data base library.

Chu, BeJijamin. Laser Light Scattering. New York: Academic Press, 1974. Series: Quantum electronics--principles and applications.

Eichler, H. J ., P. Gunter, and D.W. Poh!. Laser-induced dynamic gratings. Berlin:Springer-Verlag,1986. Series title: Springer Series in Optical Sciences; v. 50.

Electro-optics/laser international '80 UK, Brigh- ton, 25-27 March 1980: conference proceedings. H.G. Jerrard, ed. Conference organized by Kiver Communications Ltd. Guildford, Surrey, England: IPC Science and Technology Press, 1980.

Fundamentals of laser interactions II: proceed- ings of the fourth meeting on laser phenomena held at the Bundessportheim in Obergurgl, Aus- tria, 26 February-4 March 1989. F. Ehlotzky, ed. Berlin: Springer-V erlag, 1989. Series: Lecture notes in physics; 339.

ICALEO Technical Digest Eighty-Three; Vo!' 36. !bledo, OH: Laser Institute,1983.

ICALEO Materials Processing Eighty-Three: Proceedings, Vo!' 38. !bledo, OH: Laser Insti- tute,I984.

ICALEO Holography & Information Processing Eighty-Three: Proceedings,Vo!. 41. Toledo, OH: Laser Institute,I984.

Industrial applications oflaser technology: April 19-22,1983, Geneva Switzerland.William F. Fagan, chair,ed. Bellingham, WA: SPIE--the International Society for Optical Engineering, 1983. Series title: Proceedings ofSPIE--the Inter- national Society for Optical Engineering; v. 398.

The Industrial Laser Annual Handbook. David Belforte, Morris Levitt, editors; managing editor, Laureen Belleville. Tulsa, OK : PennWell Books, 1987. Series: SPIE; v. 919.

Industrial Laser Interferometry: 14-16 January, 1987, Los Angeles, California. Ryszard J . Pryputniewicz, chair/editor; sponsored by SPIE--the Inter- national Society for Optical Engineering, in coopera- tion with Center for Applied Optics, Uni- versity of Alabama in Huntsville ... ret a!.). Bellingham,

WA., USA : SPIE, 1987. Series: Proceedings of SPIE--the International Society for Optical Engi- neering.

Interferometric Metrology: 20-21 August 1987, San Diego, California. N.A. Massie, editor; spon· sored by SPIE, the International Society for Opti- cal Engineering; cooperating organizations, Applied Optics LaboratorylNew Mexico State University ... ret al.). Bellingham, WA., USA: SPIE, 1988. Series: Critical reviews of optical sci- ence and technology, SPIE; v. 816.

International Congress on Applications of Lasers and Electro-optics,1987: San Diego, California. Proceedings of the international conference on optical meth- ods in flow and particle diagnostics. Edited by Warren H. Stevenson; sponsored ... by the Laser Institute of America. !bledo, OH: Laser

Institute of America, 1988. Series: LIA Contents: v. [1) Medicine & biology - v. [2) Maters process- ing -- v. [3) Inspection, measurement & control. Laser diagnostics & photochemistry -- v. [4) Opti- cal communications & information processing. Imaging & display technology.

International Symposium on Gas-flow and Chemical Lasers (4th: 1982: Stresa, Italy). Gas flow and chem- ical lasers. Michele Onorato, ed. New York: Plenum Press,1984.

Kaminow, Ivan P. Laser devices and applications. Ivan P. Kaminow and Anthony E. Siegman, eds. New York: IEEE Press,1973. Series title: IEEE Press se- lected reprint series.

Kock, Winston E. Engineering applications of lasers and holography. New York: Plenum Press, 1975. Se- ries title: Optical physics and engineer- ing.

_ _ _ . Lasers and holography; an introduction to co- herent optics. 1st ed. Garden City, N.Y: Dou- bleday, 1969. Series title: Science study series ; [S62) .

_ _ _ . Lasers & holography: an introduction to coher- ent optics. 2nd enl. ed. New York: Dover Publications, 1981.

Laser applications to optics and spectroscopy: based on lectures of the July 8-20, 1973, Summer School, Crystal Mountain,Wash.

The Laser Marketplace in 1990: a seminar exam- in- ing recent trends and directions in the world- wide market for lasers. Morris R. Levitt, Gary T. Forrest, editors; organized by Laser Focus World in coopera- tion with SPIE--the International Soci- ety for Opti- cal Engineering. Bellingham, WA., USA: SPIE--the International Society for Optical Engineering, 1990.

Laser Measurements. !bledo, OH: Laser Insti- tute,I985.

Laser recording and information handling tech- nol- ogy. Proceedings ofa seminar held August 21- 22, 1974, San Diego, CA. Leo Beiser, ed. Palos Verdes Es- tates, CA: Society of Photo-Optical Instrumentation Engineers,1975. Series title: Society of Photo-optical Instrumentation Engi- neers Seminar proceedings; 53.

Lasers and holographic data processing. N.G. Basov, ed. Translated from the Russian by P.S. Ivanov. Mos- cow : Mir Publishers, 1984. Series title: Advances in science and technology in the USSR. Technology se- ries.

Lasers and optical radiation. Geneva: World Health Organization; Albany, N.Y.: WHO Publi- cations Centre USA [distributor),1982. Series: Environmen- tal health criteria; 23.

Laser Window and Mirror Materials. Compiled by G. C. Battle, !bm Connolly, and Anne M. Kee- see; with a pref. by Charles S. Sahagian. New York: IFIlPle- num,1977. Series: Solid State Phys- ics Literature Guides; v. 9.

Lyons, Harold. Lasers, quantum electronics, holog- raphy: part 1: Introduction to lasers: Engi- neering 823.1: a five-day short course, July 7-11, 1975: lec- ture notes. Harold Lyons, coord. Los Angeles: Univer- sity of California, University Extension, 1975.

_____ . Lasers, quantum electrorucs, holography: part I, Introduction to lasell!: Engineering 823.1, June 17-21, 1974 : lecture notes. Harold Lyons, coord. Los Angeles: Univen;ity of California, University Extension, 1974.

The Marketplace for Industrial Lasers(no fur· ther bibliog infonnation provided)

Menzel, R. Fingerprint Detection with Lasell!. New York: Marcel Dekker, 1980.

National Conference on Measurements of Laser Emissions for Regulatory Purposes, Rockville, MD., 1974. National Conference on Measurements of Laser Emissions for Regulatory Pur· poses: proceedings of a conference held in Rockville, Maryland, June 4-7, 1974. Edited by Robert H. James; cosponsored by National Bureau of Standards and U.S. Dept. of Health, Education, and Welfare, Public Health Service, Food and Drug Administration, Bureau of Radiological Health. Rockville, MD.: The Bureau; Washington: for sale by the Supt. of Docs., U.S. Govt. Print. Off., 1976 i.e. 1977.

Nonlinear optical beam manipUlation, beam combining, and atmospheric propagation: 11· 14 January 1988, Los Angeles, California. Rob· ert A. Fisher, chair/editor; sponsored by SPIE ·· the International Society for Optical Engineer- ing. Bellingham, WA.: SPIE, 1988.

Tarasov, L. V. Laser age in optics. Translated from the Russian by V. Kisin. Moscow: Mir Pub- lishers, 1981.

Trolinger, J. D. Laser applications in flow diag· nostics. J.J. Ginoux, ed. Neuilly·sur·Seine, France: North Atlantic Treaty Organization, Advisory Group for Aerospace Research and Development,1988.

_____. Laser instrumentation for flow field diagnostics. S. M. Bogdonoff, ed. Neuilly·sur· Seine, France: North Atlantic Treaty Organiza· tion, Advisory Group for Aerospace Research and Development, 1974. Series title: AGAR· Dograph; no. 186.

Ultrashort laser pulses and applications. Edited by W. Kaiser; with contributions by D.H. Auston ... [et al.]. Berlin: Springer·Verlag, [n.d.]. Series: Topics in applied physics; v. 18.

Lasers—dissertations

Cuendet, George Joseph. Optical path differ· ences induced by absorption of laser energy in beam tubes with an axial flow, due to diffusive and acoustic effects. 1983.

Frehlich, Rodney George. Laser propagation in random media. 1982.

Kachen, George Ivan Jr.. Minimization of unwanted light in laser scattering experiments. 1965.

Sandstrom, Richard Lynne. An experimental study of optical phase fluctuations in a random refractive field,1979.

This bibliography was compiled by Elizabeth Yerkes of Lyme, New Hampshire, who achieved order from chaos armed only with a Macintosh and The Chicago Manual of Style.

This Bibliography was compiled by Elizabeth Yerkes

of Hanover, NH USA

16

Glossary

Absorption Hologram: A hologram formed in a material which acquires a certain density in response to exposure. When the hologram is illuminated, part of the light which is not absorbed is diffracted into forming the image.

Achromatic: Free of color. Black and white. In optical systems, the term is used to describe lenses which correct for chromatic aberration.

Acoustical Holography: The making of holograms by using sound waves.

Additive Color Mixing: Means by which two or more frequencies are combined by superimposition to create more colors.

Ambient Light: Light present in the immediate environment. In holographic display, often used to describe background light that is not part of the hologram illumination and may interfere with the viewing of the image.

Amplitude: The maximum value of the displacement of a point on a wave front from its mean value. Graphically, t h e height or depth of the crest or trough of a wave from its zero point.

Amplitude Hologram: A hologram by which information is stored as variations in transmittance. Also called absorption hologram

Antihalation Backing (AH): A dark material placed on the back surface of a plate or film to prevent unwanted light from striking the emulsion. Useful to prevent the formation of "Newton Rings" in the hologram. Only to be used with transmission holograms.

Argon Laser: A laser which operates when argon gas is ionized and controlled by a magnetic field. Produces several blue and green frequencies.

Artistic Holography: Holograms cre- ated for the purpose of being seen and whose value derives at least in part from the image presented.

Astigmatism: An aberration caused by the horizontal and vertical aspects of an image forming in different planes.

Bandwidth: The range of frequencies over which a given instrument will operate.

Beamsplitter: An optical component which divides a beam into two or more separate beams. A 50:50 beamsplitter produces two beams of approximately equal intensities. A 90: 10 beamsplitter transmits approximately 90% of the incident beam and reflects 10% into the second beam.

Benton Hologram: Another term for rainbow hologram. Named for its inventor, Steve Benton. A hologram produced by reducing vertical information in order to correct for image dispersion.

Biconcave: A lens which has both faces curving inward. A type of negative lens.

Biconvex: A lens which has both faces curving outward. A type of positive lens

Bleach: In holographic processing, a chemical used to change an absorption hologram into a phase hologram in order to improve efficiency (brightness).

Bragg Diffraction (Bragg's Law): Diffraction which is reinforced by reflection by a series of regularly spaced planes which correspond to a certain wavelength and angular orientation. The angle at which this reinforcement occurs is Bragg's angle.

BRH: Bureau of Radiological Health. U.S. government agency responsible for setting laser safety standards.

Brightness: A subjective term describing the amount of light perceived.

Cavity: Another name for optical cavity or laser cavity.

Chromatic Aberration: Lens or hologram irregularity due to the shifting of image position for each frequency. I f severe enough, the image will appear to blur due to the lack of registration of the colors.

Coherence Length: The maximum path length difference (between the reference beam and the object beam) in a holographic set-up that can be used and still obtain a clear, bright hologram. Coherence length depends on the type of laser used and how it is made. An etalon will increase the coherence length of a laser to about 30 feet.

Coherent Light: Light which is of the same frequency and vibrating in phase. The laser produces coherent light.

Collimated Light: Light which forms a parallel beam and neither converges nor diverges. Also referred to as collimated beam.

Collimator: A device used to produce collimated light by positioning a light source at the focal point of a lens or parabolic mirror. Such a device is called a collimating lens or collimating mirror, respectively.

Color Spread: The area over which a spectrum is dispersed.

Computer Generated Hologram: A synthetic hologram produced using a computer plotter. The binary structure is produced on a large scale and then photographically reduced into a given medium. The technique allows the production of impossible or nonexistent 3-dimensional forms.

Concave Lens: A lens with an inwardly curving surface which causes light to diverge. See also Negative lens.

Concave Mirror: A mirror with an inwardly curving surface which causes light to converge.

Constructive Interference: Coherent wave fronts of the same frequency are superimposed, and their instantaneous amplitudes add up to a greater amplitude than that of the component waves.

Continuous Wave Laser: A laser which emits a beam which does not vary overtime.

Convergence: The optical bending of light rays toward each other, as by a convex lens or concave mirror.

Convex Lens: A lens with an outwardly curving surface which causes light to converge, usually to a focal point.

Copy Hologram: Another term for image plane hologram or any second generation hologram produced from a master hologram. A contact copy is produced by placing the plate in contact with the origi- nal.

Copy Plate: Another term for copy hologram. Usually refers to the plate before it is exposed.

Cross Hologram: Another name for the type of holographic stereogram which incorporates the advantages of rainbow holography. Named for Lloyd Cross.

Cross Talk: The phenomenon of spurious images formed by color holograms when an interference pattern formed by one color also reconstructs an image in another color.

Cylindrical Mirror/Lens: An optical component which causes light to focus as a slit or line by passing through or reflecting from a surface curved in one dimension.

Denisyuk Hologram: Another name for single beam reflection hologram. Named for its inventor, Y. N. Denisyuk.

Density: The amount of opacity or darkness of a medium.

Depth of Field: The area within which satisfactory resolution of an image can be obtained. Also, in holography, used to describe the area within which any image can be formed, due to the constraints of coherence length.

Developer: A chemical solution which changes the latent image of a photographic image or holographic interference pattern

(silver salts) into black metallic silver. The term development usually refers to the degree of effect of the developer or the cause of the amount of density.

Dichromated Gelatin (DCG): A light-sensitive gelatin made up of a solution of dichromate compound, usually ammonium dichromate, in the presence of a gelatin substrate. Exposure results in the cross- linking of gelatin molecules with each other.

Diffraction: The change in direction of a wave front at the edges of an aperture, caused by the wave nature of light. Diffrac- tion is not the same process as reflection or refraction.

Diffraction Efficiency: In a hologram, the percentage of incident illumination light diffracted into forming the image. The greater the diffraction efficiency, the brighter the image will appear in a given light.

Diffraction Grating: A holographic diffraction grating is a hologram formed by the interference of two or more beams of pure, undiffused laser light.

Diffuse Reflector: An object that scatters illumination striking it. Most objects are diffuse reflectors.

Divergence: The bending of light rays away from each other, usually by concave lens or convex mirror, so that the light spreads out. Light will also diverge with a convex lens or concave mirror after it passes through the focal point.

Double Exposure: The formation of two holograms on the same recording medium. Used to cause either overlapping images or two discrete images to appear under different conditions.

Electromagnetic Radiation: Radiation emitted from vibrating, charged particles, all of which travels through space at the speed of light. Visible light is only a small part of the entire electromagnetic spectrum.

Embossed Hologram: A hologram copy made by pressing a metal surface relief master hologram into plastic film, or by using the master hologram in a mold.

Emulsion: A suspension of light sensitive silver salts (e.g. silver bromide) in gelatin, usually coated onto glass, polyester film, or by using the master hologram in a mold.

Exposure: The act or time of allowing light to impinge upon the emulsion.

Film Plane: The plane at which the recording material is located.

Fixer: A chemical solution which removes the unexposed silver salts from the emulsion to desensitize the emulsion after development.   .

f-number: The ratio of the focal length of a lens or curved mirror to its diameter.

Focal Length: The distance from the center plane of a lens or curved mirror to a position where a collimated beam is focused to a point. A lens with a negative focal length (a concave lens) appears to have a focus upstream from the lens, while a lens with a positive focal length (a convex lens) has a focus downstream from the lens.

Focused Image Hologram: Any hologram in which the image appears on the surface of the hologram or seems to intersect the surface of the hologram.

Fog: The darkening or exposing of film by inadvertently allowing ambient light to strike it. In holography, a fogged plate reduces fringe contrast, resulting in a less efficient image.

Fourier Transform Hologram: A hologram made using a reference beam diverging from a point at the same distance from the recording plate as the object. Also called Fraunhofer Hologram

Frequency: The number of crests of waves that pass a fixed point in a given unit of time.

Fresnel Hologram: Another name for the common hologram. Defined as a hologram formed with an object located close to the recording medium.

Fringe: An individual interference band, made up of one cycle of constructive and destructive interference.

Front Surface Mirror: A mirror with the reflecting surface on the front. Conventional mirrors have their reflecting surfaces on the back of a piece of glass and are not useful for holography as the front surface produces a "ghost" reflection.

Gabor Hologram: An in-line hologram of the type invented by Dennis Gabor.

Gas Laser: A laser such as a Helium-Neon laser, in which the lasing medium is a gas.

Grating (also Diffraction Grating): A pattern of very fine lines of equal spacing, usually on the order of a few microns apart. A diffraction grating can redirect light and break white light into its component colors like a prism.

H-1: The first hologram made in the process of making a master hologram. An H-1 has an image only viewable in laser light.

H-2: The second hologram made in the process of making master hologram. A master hologram is usually an H-2.

Helium-Neon (HeNe): The most common lasing material, which produces a continuous red beam at 632.8 nm.

HOE: Holographic Optical Element. A hologram which may be used to act as a lens, mirror, or a complex optical component.

Hologram: An interference pattern formed as a result of reference light en-coun- tering light scattered by an object and stored as such on a light sensitive emulsion.

Holographic Movie: The animation of a 3-dimensional holographic image by presentation of numerous holograms in rapid sequence in much the same way motion pic- ture film operates. Unlike conventional cin- ema, it is only with extreme difficulty that the image can be projected, and true holo- graphic movies are still very experimental. The term is often used, incorrectly, for holographic stereograms.

Holographic Stereogram: A holo-gram made by filming numerous angles of view of a scene and then storing the frames holographically. Each eye views a different frame, displaced so as to result in the illusion of a stereoscopic image. Also called a multiplex or integral hologram.

Image Plane Hologram: A second generation hologram formed by positioning a light sensitive plate in the plane of an image formed by a master hologram.

Incandescent Light: Light formed when an electric current passes through a resistant metal wire, usually situated in a vacuum bulb.

Incoherent Light: Light which is emit-ted with randomly varying phase and a mix of colors. Light from ordinary souroes such as light bulbs is incoherent; laser light is coherent.

Index of Refraction: The ratio of the velocity of light in air to the velocity of light in a refractive material for a given wavelength.

Infrared: That part of the spectrum characterized by wavelengths somewhat longer than those of red light, which are not visible to the eye yet are often perceived as heat. Covers the spectrum from about 750 nm to 1000 micrometers.

In-Line Hologram: A Gabor hologram. Made by positioning the object and reference light along the same axis, resulting in a configuration practical only for making holograms oftransparencies.

In Phase: The relationship of two waves of the same frequency when they travel through their maximum and mini- mum values simultaneously and are also polarized identically. Holograms must be made by waves which remain in phase dur- ing the course of an exposure.

Interference: The result of superim-posing two or more waves. The waves oscillate between negative and positive values, so when two waves are superimposed the positive values reinforce each other while negative values cancel out positive values.

Interference Pattern: A stationary pattern of interference that results when light waves are superimposed.

Interferometer: A device that utilizes interference of light to measure changes in systems with extreme accuracy. An inter-fer- ometer can be used to test the stability of holographic systems.

Ion: An atom or molecule which has gained or lost an electron so that it acquires a negative or positive charge.

Ion Laser: A laser within which stimu-lated emission occurs as a result of energy changes between two levels of an ion. Argon and krypton are the two most common types of ion laser.

Krypton Laser: An ion laser that pro-duces many frequencies which appear over a large part of the spectrum. The most common lines are blue, green, yellow, as well as a very strong red frequency.

LASER: From the acronym for "Light Amplification by Stimulated Emission of Radiation." A laser is usually in the form of a light-amplifying medium placed between two mirrors. Light not perfectly aligned with the mirrors escapes out the sides, but light perfectly aligned will be amplified. One mirror is made partially transparent. The result is an amplified beam of light that emerges through the partially transparent mirror.

Latent Image: The image or pattern stored in an emulsion before it is developed into a visible image.

Latent Image Decay: A condition that is common to fine grained silver emulsions, including the types used for holography. The decay occurs if the material is not processed soon after exposure, resulting in a lower density.

Light Meter: Any device used to sense and measure light. Usually used to sense intensity in order to determine exposure.

Line Spacing: The distance between individual interference fringes in a diffraction grating.

Lippman Hologram: Another name for reflection hologram.

Liquid Gate: A liquid-tight glass-sided plate holder in which the plate is exposed, processed and viewed with the cell filled with liquid throughout.

Liquid Lens: A lens formed by filling a shaped glass or acrylic tank with liquid such as mineral oil. A liquid lens is often used in a system for making Cross holograms.

Master Hologram: The original H-2 hologram, from which copy holograms are made.

Mode: A laser can oscillate in a number of different modes. Spatial modes are the different repetitive patterns in which light can zigzag between the two laser mirrors. The optimum spatial mode for holography has no zigzag at all (called the TEMOO mode), resulting in a single emitted beam of light. Higher modes can result in a beam that gives a donut-shaped spot, or a clo-ver- leaf pattern. Longitudinal modes are the different wavelengths that are simultaneously emitted by a laser. A single-mode laser produces a single wavelength of light, and can only have a single spatial mode and a single longitudinal mode.

Moire Pattern: A highly visible type of (interference pattern formed when gratings, screens, or regularly spaced patterns are superimposed upon each other.

Monochromatic: Light or other radiation with one single frequency or wavelength. Since no light is perfectly monochromatic, the term is used loosely to describe any light of a single color over a very narrow band of wavelengths.

Motion: The effects of an object or holographic system not remaining rigidly fixed during exposure.

Multichannel Hologram: A hologram formed with two or more separate reference beams or angles.

Multiple Exposure: More than one exposure occurring on the same plate or film.

Multiplex: Another name for holographic stereogram.

NAB: A holographic plate without an antihalation backing. Also called an unbacked plate.

Negative Lens: A lens characterized by a concave surface which causes light to diverge. A negative lens has a negative focal length.

Newton's Rings: The series of rings or bands which appear due to interference between two nearly parallel surfaces. These rings often form as a result of light interacting between the front and back surfaces of a holographic plate.

Node: The part of a vibrating wave that is not moving - zero point. An antinode is a point on the wave of maximum displacement from the zero point.

Noise: Any unwanted light scattering by components in a holographic set-up or by particles in a holographic recording medium.

Object Beam: The light beam in a holographic set-up which illuminates the object. Also, the light reflected from or transmitted by the object which is recorded in the hologram.

Off-Axis: The type of hologram invented by Emmet Leith and Juris Upatnieks whereby object and reference beams approach the holographic plate at different angles.

On-Axis: Hologram formed with object and reference beams originating along the same axis. Also called an in-line or Gabor hologram.

Open Aperture: A transmission image plane hologram viewable in white light and characterized by both vertical and horizontal parallax and usually a brilliant white image.

Optical Cavity: The space between the two mirrors in a laser. The tube is located within the optical cavity.

Optical Component: An optical device consisting of the optics (lens, mirror, etc.) and a mount used to affix it to a vibration isolation table.

Optics: Those devices which change or manipulate light, including lenses, mirrors, beamsplitters, filters, etc. Also the science of electromagnetic radiation, its effects, and the phenomenon of vision.

Orthoscopic: Having the "right" appearance. Orthoscopic image has the correct appearance, whereas a "pseudoscopic" image appears to have its depth inverted.

Oscillator: Any device that converts energy into an alternating electromagnetic field, usually of constant period.

Overexposure: Improper resulting from too much light or light reaching the plate or film for too long.

Parabolic Mirror: A mirror with a surface curved in the shape of a parabola. Used as a telescope mirror in astronomy or as a collimating mirror in holography.

Parallax: The difference between two different views of an object, obtained by changing viewing position.

Period: The time required for a wave to go through one complete cycle. The period of a typical light wave is about one trillionth of a second.

Phase: A wave oscillates from a positive value to a negative value and back to positive. The phase of a wave relates to where it is in its oscillation cycle at a particular moment. Usually only the phase difference between two waves is important.

Phase Hologram: A hologram which diffracts light by delaying the phase of certain portions of the lightwave, rather than absorbing certain portions. Bleached silver halide holograms, DCG holograms, and surface relief holograms are phase holograms, while unbleached silver halide holograms are amplitude holograms.

Phase Shift: The amount by which the phase of one light beam is delayed or advanced relative to another light beam.

Photochemistry: The branch of chemistry dealing with the effects of light on chemical reactions.

Photon: The smallest unit or quantum of electromagnetic energy known today.

Photopolymer: A material which "polymerizes" where it is exposed to light. Photopolymers are usually partially solidified plastics which finish solidifying when exposed.

Photoresist: A chemical substance made insoluable by exposure to light (usually ultraviolet). Although most often used to manufacture microcircuits, photoresist can be used to make holograms.

Pinhole: The small hole used to pass focused light from the objective in a spatial filter.

Plane Hologram: A hologram for which fringes are large with respect to the thick- ness of the emulsion, so that interference is mostly stored on the surface of the holo- gram.

Plano-Concave: A lens which has one concave surface and one flat surface.

Plano-Convex: A lens which has one convex surface and one flat surface.

Plateholder or Platen: Any device which holds a holographic plate or film in place during the exposure.

Polarization: The restriction of light or other radiation to vibration in only one plane.

Population Inversion: A condition whereby more atoms are in the excited state than in the ground state, resulting in the predominance of stimulated emission.

Positive Lens: A lens with an outwardly curving surface which causes light to converge. Also known as convex lens.

Processing: The entire chemical sequence, from development to final drying of the hologram.

Pseudocolor: The production of colors in a hologram which are not related to the true colors of the original objects. Usually used in connection with multicolor holograms.

Pseudoscopic: The opposite of orthoscopic. An image whose parallax is reversed.

Pulsed Hologram: A hologram produced with the short burst of light from the pulsed laser. May be used to make holograms of live subjects.

Pulsed Laser: A laser which emits radiation in a wave of short bursts and is inactive between bursts.

Quantum: The smallest amount that the energy of a wave may be divided into.

Rainbow Hologram: A white light viewable hologram with colors which shift through the spectrum as the hologram or the viewing angle is tilted. A rainbow hologram has no vertical parallax.

Real Image: An image that is formed in such a way that it actually comes to a focus. A real image is one that forms down- stream from a lens; a virtual image appears to form upstream from a lens.

Real Time Holography: A technique whereby a holographic image is superim- posed over a real object in order to observe interference fringes generated by minute changes between the two.

Reconstruction Beam: Light directed at the finished hologram from which the object wave front will be recre- ated.

Recording Material: Any substance which may be used to record the interference pattern of the hologram.

Reference Angle: The angle at which the reference beam strikes the plate, usu- ally measured in degrees from the plate surface.

Reference Beam: The unmodulated, pure laser light directed at the plate to interfere with the object light.

Reflection Hologram: A hologram made by allowing reference and object light to impinge on opposite sides of the plate. The finished hologram is viewed by allowing light to reflect from it to the observer.

Refraction: The bending of light which occurs when it passes from a medium of one refractive index to that of another. In a phase hologram, refraction causes a "phase delay" which corresponds to the original phase difference between the two stored wave fronts.

Refractive Index: Same as index of refraction.

Resolution: The ability of a film or an optical system to distinguish between two closely spaced points. Film resolution is usually expressed in terms of how many closely spaced lines per millimeter the film can record. Holographic films must be capable of high resolving capability since the in-

terference fringes are often extremely small and closely spaced.

Resonance: A large amount of vibration in a system which is caused by a small stimulus with approximately the same period as the natural vibration period of the large system.

Resonant Cavity: Another name for optical cavity or laser cavity.

Scatter: Unwanted light which interferes with the making of a good quality hologram.

Settling Time: A period of time between the loading of the plate and the exposure in order to allow ambient vibra- tions time to dampen.

Set-up: The configuration of optical components used to produce a given hologram.

Shadowgram: A hologram made by deliberately moving an object during an exposure, or by using an inherently unsta- ble object, in order to produce a 3-dimen- sional "hole" or shadow where the object was once located.

Shutter: The device used to block the laser beam and then allow it to pass unobstructed for the desired exposure time.

Silver halide: The type of recording material which consists of light-sensitive silver particles suspended in gelatin.

Single beam hologram: A hologram made with one beam which acts as both reference and object illumination beam.

Slab Table: An optics table which uses a concrete slab as part of its inertial mass.

Slit Optics: Any optical device which causes light to be propagated into a line. Usually formed by light interacting with a cylindrical surface.

Solid State Laser: A laser which uses a solid material, such as ruby, as its lasing medium.

Space: The area between objects. In holography, the area between and including objects .

Spatial Filtering: The act of "cleaning up" the light of the laser beam by causing it to focus through a tiny aperture. Only the pure light can focus at the desired point, eliminating the effects of dust, optical surface scratches, etc.

Spatial Frequency: Often used with regard to line spacing in diffraction gratings. The spatial frequency is the reciprocal of line spacing, generally expressed in cycles per millimeter. See also resolution.

Speckle: The grainy appearance of an object, or a holographic image, viewed under laser light. It is caused by light reflecting from minute areas of the object and interfering with itself.

Spectral Reflection: Any reflection from a smooth, polished surface, such as a mirror. Also called specular reflection.

Splitbeam: The act of separating a beam of laser light into two components to separately control the action of reference and object illumination.

Squeegee: A device or action used to remove excess water from the emulsion to facilitate drying.

Stability: The requirement for holographic optical systems to remain motionless during an exposure.

Standing Wave: The result of superimposing two or more waves moving in differ- ent directions but having the same wavelength. An interference pattern is a slice through a standing wave pattern; and a hologram is a photograph of an interfer- ence pattern.

Stereogram: An image which creates a 3-dimensional illusion by presenting a different view of an object to each eye.

Stimulated Emission: Radiation produced by incoming radiation of the same phase, amplitude, and frequency.

Stop Bath: The chemical bath immediately following the developer which causes the developer to cease action.

Temoo: The lowest mode of a laser, characterized by a beam which is spatially coherent across the diameter of the beam.

Temporal Coherence: Coherence over time. The degree to which waves will remain coherent over time and distance.

Test Strip: A means of visually determining the correct exposure by making a series of individual exposures of varying times on the same plate. The proper time is determined by selecting the strip which yields the brightest or cleanest image.

Thermoplastic Film: A recording material which works due to the effects of electrostatic forces and heat to produce a deformation corresponding to the interfer- ence pattern exposed.

360-Degree Hologram.: A hologram made by exposing recording material which completely surrounds an object.

Transfer Mirror: A mirror which redirects light from the laser toward the desired working area on the optics table.

Transmission Hologram: Any hologram viewed by passing light through it, toward the viewing side. Transmission holograms are made by allowing both object and reference light to impinge on the same side of the plate.

Transmittance: The proportion of light transmitted by a medium to that which is incident upon it.

Triethanolomine: A chemical used to change the thickness of the emulsion to produce different color playback, usually with reflection holograms.

Ultraviolet: An invisible part of the spectrum characterized by wavelengths somewhat shorter than violet (approx. 100-400 nm.).

Unbacked Plate (NAB): A holo- graphic plate without an antihalation back- ing. Essential for reflection holograms.

Variable Beamsplitter (VBS): A beamsplitter whereby the ratio of transmit- ted to reflected beam changes as the beam intercepts the component at different points.

Vibration Isolation: The practice of removing a system from the effects of ambient vibrations which may induce changes, particularly in optical systems. Vibration isolation must be used in making a hologram to prevent the movement of interference fringes during an exposure.

Virtual Image: An image which appears "upstream" from a lens or holo- gram. A real image becomes visible when a piece of paper is placed in its location, but a virtual image is not accessible.

V olume Hologram: A hologram in which the thickness of the recording material is large compared to the spacing of the interference fringes. DCG and photopolymer holograms are volume holograms; embossed holograms are "thin" or "surface" holograms.

Wave Form: The characteristic shape taken on by a wave front.

Wave Front: The surface of a propagat- ing wave, where the phase of the wave is the same everywhere on the surface. A point source produces spherical wavefronts, a collimated beam consists of plane wavefronts, and light reflected from a complicated object has a wavefront with a very complicated shape.

Wavelength: The physical distance over which the complete cycle of one wave occurs. Wavelength is inversely proportional to frequency.

White Light Transmission Hologram: Any transmission hologram which can be displayed using ordinary white light.

YAG Laser: A solid state laser using Yttrium Aluminum Garnet as the lasing material.

Zone Plate: A pattern consisting of a central spot surrounded by concentric zones, alternating opaque and transparent, the total area of each zone being equal. A zone plate is equivalent to the hologram of a point object.

DID YOU BORROW THIS COPY?

Get Your Own!

Now is the time to order your personal copy of the Holography Marketplace. This international directory is the first and only resource of its kind. You will refer to the HMP day after day, so don't you want one of your own?

Standing Order Plan

If you would like to receive the *Holography Marketplace* each year, ask to be put on our Standing Order Plan. We will ship each new edition to you upon publication.

To order by check:

Mail check or money order to:

ROSS BOOKS
HOLOGRAPHY MARKETPLACE
P.O. BOX 4340
BERKELEY, CA 94704 USA

To order by Credit Card:

Credit card customers can order by phone using Visa, Mastercard or American Express. Telephone: (800) 367 0930 or (510) 841 2474

Cost

For USA/Canada/Mexico delivery:
US$39.00 (Includes shipping costs)

For delivery outside North America:
We will accept your check drawn on any bank in the world for any one of the following amounts *(all prices include airmail shipping)*:

Europe: US$50.00/ DM100/ £33

Asia / Africa / Pacific Rim: US$55/DM110/£37

Western Hemisphere: US$45.00/DM90/£30

FAX your order to: (1) (510) 841 2695

For orders by mail or fax, fill out the information below. Include check or credit card information.

NAME: _____

COMPANY:_____

ADDRESS:_____

CITY/STATE /PROVINCE:_____

COUNTRY/POSTAL CODE:_____

PHONE/FAX:_____

Credit Card Information

American Express ❏   Mastercard ❏   Visa ❏
Card #_____
Expiration Date_____
Signature_____

Please mail me a form to list my company in the next edition of Holography Marketplace ❏

FOR
VISUAL IMPACT...

PUT YOUR DESIGNS INTO A NEW DIMENSION.

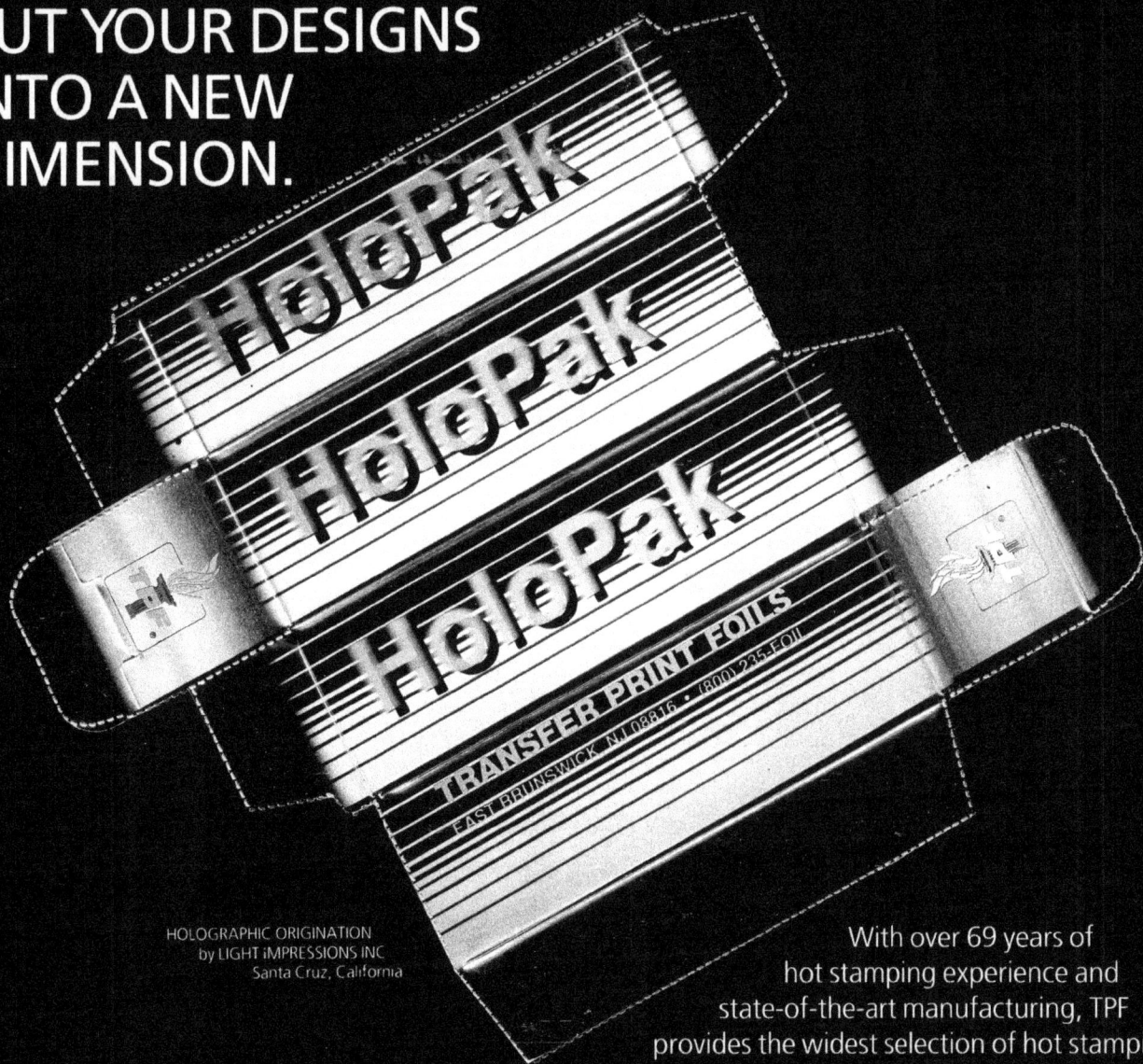

HoloPak
HoloPak
HoloPak

TRANSFER PRINT FOILS
EAST BRUNSWICK, NJ 08816 • (800) 235-FOIL

HOLOGRAPHIC ORIGINATION
by LIGHT iMPRESSIONS INC
Santa Cruz, California

With over 69 years of
hot stamping experience and
state-of-the-art manufacturing, TPF
provides the widest selection of hot stamping
foils and laminates for decorating paper and plastics.

Holographic Foils & Gratings _ Metallized Paper & Film For Packaging
Metallics _ Pearls _ Pigments _ Standard Patterns _ Custom Patterns

TPF Transfer Print Foils, Inc.

9 Cotters Lane • PO Box 538 • East Brunswick, New Jersey 08816
(908) 238-1800 • FAX (908) 651-1660 • #1-800-235-FOIL

www.ingramcontent.com/pod-product-compliance
Lightning Source LLC
Chambersburg PA
CBHW051345200326
41521CB00014B/2483